SpringerBriefs in Physics

T0210675

Cam Nguyen · Seoktae Kim

Theory, Analysis and Design of RF Interferometric Sensors

 Springer

Cam Nguyen
Texas A&M University
College Station, TX, USA
cam@ece.tamu.edu

Seoktae Kim
School of Engineering Information
and Communications University
Daejeon, South Korea
setakim@pknu.ac.kr

ISSN 2191-5423   e-ISBN 2191-5431
ISBN 978-1-4614-2022-4   e-ISBN 978-1-4614-2023-1
DOI 10.1007/978-1-4614-2023-1
Springer New York Dordrecht Heidelberg London

Library of Congress Control Number: 2011941613

Printed on acid-free paper

Springer is part of Springer Science+Business Media (www.springer.com)

Preface

One of the very fundamental implicit principles of sensing is detecting changes. As any change causes variation of the phase of electrical signals used for sensing, any technique that can rapidly and accurately detect the phase change would serve as a good candidate for sensing. RF interferometry is such an attractive sensing technique. It is basically a phase-sensitive detection process, capable of quickly resolving any measured physical quantity within a fraction of the operating wavelength.

This book is devoted to the theory, analysis and design of RF interferometric sensors using RF integrated circuits. It also presents the measurement of displacement and velocity as a way to demonstrate the sensing ability of the RF interferometry and to illustrate its many possible applications in sensing. Although the book is succinct, the material is very much self-contained and presented in a way that allows readers with an undergraduate background in electrical engineering or physics with some experiences or graduate courses in RF circuits to understand and implement the technique easily.

The book is useful for engineers, physicists and graduate students who work in sensing areas, particularly those involving radio waves. It is also useful for those involved in RF system design. It is our sincere hope that the book can serve not only as a reference for the development of RF interferometric sensors, at least as the first step in this endeavor, but also for a possible generation of innovative ideas that can benefit many existing sensing applications or be implemented for other new applications.

College Station, TX, USA   Cam Nguyen
Daejeon, South Korea   Seoktae Kim

Abstract

This book presents the theory, analysis and design of RF interferometric sensors. RF interferometric sensors are attractive for various sensing applications that require very fine resolution and accuracy as well as fast speed. The book begins with an introduction of the original optical interferometry and its RF interferometry deviation. It then presents an analysis of RF interferometers with particular focus on operation at millimeter wavelengths. Two types of RF interferometric sensor configurations, namely homodyne and double-channel homodyne, operating in Ka-band (26.5–40 GHz) are discussed in details including the system architecture, signal processing, theory, analysis, design, fabrication and test. These sensors are especially realized on planar structures using RF integrated-circuit technology for light weight, compact, and low cost. Various measurements for the designed sensors are conducted to verify the sensor design and to demonstrate the usefulness and some sensing applications of these sensors. These include displacement measurement, liquid-level gauging and velocimetry which serve as representative examples of many possible sensing applications of RF interferometric sensors. The developed millimeter-wave homodyne sensor shows that sub-millimeter resolution in the order of 0.05 mm is feasible without correcting the non-linear phase response of the sensor's quadrature mixer. The designed millimeter-wave double-channel homodyne sensor provides a better resolution of 0.01 mm or $1/840^{th}$ of the operating wavelength than the sensor employing the simple homodyne configuration. The double-channel homodyne sensor can suppress the non-linearity of the sensor's quadrature mixer. This sensor especially employs a digital quadrature mixer, constituted by a quadrature-sampling signal processing technique, enabling elimination of the conventional quadrature mixer's nonlinear phase response.

The developed RF interferometric sensors demonstrate that displacement sensing with micron resolution and accuracy, and high-resolution low-velocity measurement are feasible using millimeter-wave interferometry, which is attractive not only for displacement and velocity measurement, but also for other sensing applications requiring very fine resolution and accuracy as well as fast speed.

Keywords Interferometry • Interferometer • RF interferometer • Microwave interferometer • Millimeter-wave interferometer • RF interferometer • Interferometric sensor • Sensor • Radar • Sensing

Contents

1 Introduction 1

2 Analysis of RF Interferometer 7
   2.1 Interaction of Electromagnetic Waves with Dielectric 8
   2.2 Determination of Relative Dielectric
        Constant and Thickness 9
   2.3 Signal Analysis of RF Interferometer 12

3 RF Homodyne Interferometric Sensor 15
   3.1 System Configuration and Principle 15
   3.2 Phase Unwrapping Signal Processing 18
   3.3 System Fabrication 20
   3.4 Displacement Measurement and Liquid-Level Gauging 22
   3.5 Analysis of Error Contributed by Quadrature Mixer 25
        3.5.1 Quadrature Mixer Transfer Function 25
        3.5.2 I/Q Error Correction Algorithm 28
        3.5.3 Worst-Case Error Analysis 31
   3.6 Summary 33

4 Double-Channel Homodyne Interferometric Sensor 35
   4.1 System Configuration and Principle 36
        4.1.1 Displacement Measurement 38
        4.1.2 Doppler Velocimetry 39
   4.2 Signal Processing 40
        4.2.1 Phase-Difference Detection for Displacement
               Measurement 40
        4.2.2 Doppler-Frequency Estimation for Velocity
               Measurement 42
   4.3 System Fabrication and Test 46
        4.3.1 Displacement Measurement Results 48
        4.3.2 Velocity Measurement Result 51
   4.4 Summary 54

**5 Consideration of Frequency Stability of RF Signal
    Source for RF Interferometer** 55
    5.1 Theoretical Analysis of Phase-Noise Effect
         on Interferometric Measurements..................................... 55
    5.2 Phase Noise Estimation 61

6 Summary, Conclusion and Applications 65

References 69

Index 73

Chapter 1
Introduction

In general, interferometry is a scientific technique, as it literally implies, to interfere or correlate two or more signals to form a physically observable measure, like a fringe pattern in optical interferometry or electrical signals in most radio interferometry, from which useful information can be inferred. The history of interferometry dates back to 1887 when American physicist A. A. Michelson first demonstrated optical interferometer experimentally to measure the speed of light, which later became the foundation of Einstein's Theory of Relativity.

The basic building blocks of the Michelson interferometer, which is composed of a coherent light source, two mirrors, a beam splitter, and a detector, are shown in Fig. 1.1. The Michelson interferometer works on the principle that the coherent light wave split by the beam splitter forms constructive or destructive light intensity variation, interference fringe, according to the phase relationship between the two waves split when they are interfered. In the interferometer, the light source is divided into two waves. One is used as a reference wave traveling along the path indicated by the dotted-line arrow in Fig. 1.1. The other serves as a measurement wave whose traveling path is depicted by the solid-line arrow. Interfering these two waves results in a fringe image that is used for interferometric measurement. Michelson later extended his experiment into the study of spectral lines, measurement of the standard meter, and even the measurement of the angular diameter of stars [1]. Since the first interferometer was devised, many different forms of interferometers have been investigated with different frequency sources. The introduction of laser has played great role in the development of optical interferometry. Today, interferometry applies to various areas such as measurement techniques as well as medicine and biology [2, 3].

Although major achievements of interferometry originate from optical interferometry, radio frequency (RF) interferometry, using spectrums in the RF range,[1] has been investigated enormously in areas such as radio astronomy and astrometry,

---

[1] The RF range here is loosely defined as including all the frequencies from 1 MHz to 300 GHz (millimeter waves).

C. Nguyen and S. Kim, *Theory, Analysis and Design of RF Interferometric Sensors*, SpringerBriefs in Physics, DOI 10.1007/978-1-4614-2023-1_1, © Springer Science+Business Media, LLC 2012

Fig. 1.1 Michelson
interferometer

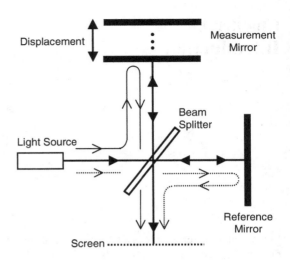

Fig. 1.2 Typical schematic
of RF interferometer

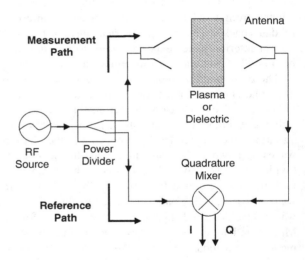

plasma diagnostics, nondestructive material evaluation, and sensing applications. The RF interferometer has many similarities to radar in the aspects of structure and principle. The RF interferometer is typically used for short distances, while radar is mainly used for long distance applications with some short-distance exceptions such as at 60 GHz. Therefore, the RF interferometer can be regarded as a kind of coherent radar in terms of radar terminology. As an example of the RF interferometer, Fig. 1.2 shows a typical schematic diagram of the RF interferometer to measure electron density in a plasma chamber or to evaluate complex permittivity of a dielectric medium located between two antennas. The structure of the RF interferometer shown in Fig. 1.2 is analogous to the Mach-Zehnder interferometer [4] used in optics, except for the RF components used to build the system.

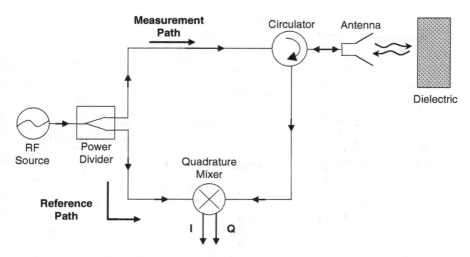

Fig. 1.3 RF interferometer employing one antenna to measure reflected wave

In the RF interferometer depicted in Fig. 1.2, a RF wave is divided by the power divider, which can also be replaced with a directional coupler as appropriate, into two waves traveling in two different paths, namely the measurement and reference path. A dielectric medium with properties to be measured is placed in the measurement path. The wave in the measurement path carries information to be measured, such as the electron density of plasma or the material properties of dielectric, after passing through the medium. This wave then interferes with the wave in the reference path by means of the quadrature mixer, resulting in in-phase (I) and quadrature (Q) signals from which any physical quantity relating to the properties of dielectric is extraced. It is also possible to constitute interferometer employing only one antenna, as shown in Fig. 1.3, detecting and characterizing the reflected wave instead of the transmitted wave. In view of radar engineering, this is generally called a reflectometer or monostatic system. Due to its inherent versatility, this type of system has been used for a wide range of applications and is also adopted as the system topology for different RF interferometers discussed in this book. In the system, a circulator is used to seperate the reflected wave from the transmitted wave.

Another form of RF interferometers is the double-channel system as shown in Fig. 1.4, where either a reference or measurement wave is modulated by a quadrature upconverter to generate a single-sideband (SSB) signal which is slightly shifted in frequency by f_m with reference to the frequency f of the signal source. In this system, the output signal of the mixer is not DC or zero-IF, as previously discussed for the systems in Figs. 1.2 and 1.3. The measurement process is accomplished by comparing the phase of the modulating signal of frequency f_m to the output signal of the mixer that has a frequency of f_m and contains information on the material to be evaluated. In Fig. 1.4, the RF interferometer employing two antennas is illustrated as an example, but one antenna configuration is also possible.

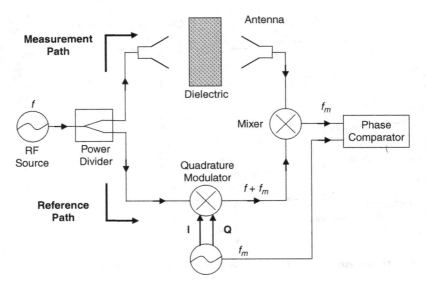

Fig. 1.4 Double-channel RF interferometer

The main advantages of this topology are its ease in avoiding the 1/f noise generated by the semiconductor components used and elimination of the I/Q error of the quadrature mixer, as will be explained in Chap. 3.

Optical interferometry, depending on the number of light sources used, is traditionally classified as a homodyne or heterodyne interferometer. In a homodyne system, the two interfered waves are produced by the same light source. On the other hand, in a heterodyne system, the two interfered waves have different frequencies coming from two different light sources. Modern advances in optical interferometry give rise to further classifications, besides the number of light sources, based on the parameters to describe the interferometers [4, 5]. However, strict classifications are not found in RF interferometers. The nomenclature of homodyne and double-channel homodyne interferometric sensors is adopted throughout this book, relying on the number of junctions to divide and combine RF waves in the systems, as proposed in [6].

RF sensors at millimeter-wave frequencies (30–300 GHz), or millimeter-wave sensors, have been widely investigated for various sensing applications from military to commercial and medical implementations such as contraband detection, process control, automobile collision avoidance, early cancer detection, etc. With the advances in high-speed solid-state electronics operating in millimeter-wave frequency range, RF sensors have been developed in a lighter, cheaper, and more compact way than ever before. Those advances make it possible to design and implement radar sensors in planar structure. Moreover, the development of milli-meter-wave integrated circuits (IC) on silicon such as millimeter-wave CMOS and BiCMOS RFICs has progressed rapidly, making feasible the development of millimeter-wave sensors incorporating RF and digital circuits in a single chip for extremely small size and low cost. The RF interferometer is a unique RF sensor capable of providing accurate measurement with very fine resolution. It is capable of resolving any measured physical quantity within a fraction of the operating wavelength. It also has a relatively faster system response time than other sensors due to the fact that they are generally operated with single-frequency sources. As such, the RF interferometer has many applications crossing many different areas from material measurement to defense applications, like mine detection, infrastructure monitoring, to medicine like cancer detection and heart-beat monitoring. RF interferometry in the microwave and millimeter-wave ranges has been used widely for non-destructive material characterization [1, 7], plasma diagnostics [8], position sensing [9, 10], velocity profile [11], cardio pulmonary [12], radio astronomy [13, 14], topography [15], meteorology [16], precision noise measurement [17], displacement measurement [18], and low-velocity measurement [19].

In this book, we present the theory, analysis and design of the RF interferometric sensors. To illustrate the design for these sensors, we particularly present the development of millimeter-wave homodyne and double-channel homodyne interferometers and demonstrate various measurements including displacement, liquid level gauging, and velocity. These millimeter wave interferometers are designed using microwave integrated circuits (MICs) and microwave monolithic integrated circuits (MMICs). Much progress in millimeter-wave sensors is found with pulse or FMCW (Frequency Modulated Continuous Wave) techniques. In this book, we also demonstrate that millimeter-wave interferometers operating at a single-frequency CW provides a more attractive solution than the pulse or FMCW radar sensor for short-range applications requiring high resolution and fast response.

The book is organized as follows: Chap. 1 gives the introduction with discussion of the original optical interferometer and RF interferometers derived from it. Chapter 2 analyzes the principle of RF interferometry through the study of relative dielectric constant and thickness of dielectric materials; Chap. 3 is devoted to an interferometric sensor with homodyne system configuration and its use for displacement measurement and liquid level gauging; Chap. 4 describes an interferometric sensor with double-channel homodyne configuration for displacement and velocity sensing; Chap. 5 includes the analysis of a phase noise effect on interferometric measurement; and finally, Chap. 6 provides the summary, conclusion and applications.

Chapter 2
Analysis of RF Interferometer

In this chapter, the principle of RF interferometry is investigated for the measurement of the permittivity and thickness of dielectric as shown in Figs. 1.2, 1.3, and 1.4 of Chap. 1, as an exemplary study. It is shown that the permittivity and thickness of dielectric can be determined from the measured phase of the reflection and transmission of plane electromagnetic waves reflected from or transmitted through the material. It should be noted that the same principle can be applied for general interferometric measurement, for example, displacement, distance and velocity measurements, by defining the relationship between the phase detected and any physical measure to be evaluated. Applications of the principle to measure the change of position of metal plate, to gage liquid level, and to estimate low velocity of a moving object are found in Chaps. 3 and 4. Signals of the measurement system to probe the phase are analyzed.

2.1 Interaction of Electromagnetic Waves with Dielectric

Figure 2.1 illustrates a geometry involved in the analysis of interferometry. It is assumed that the dielectric is located in the far field so that the incident electromagnetic wave is a plane wave. The electric field of an electromagnetic wave normally incident on the dielectric and traveling in the z direction is expressed as

$$E_i(z, t) = E_0 \exp(j\omega t - \gamma z) \tag{2.1}$$

where E_0 is the initial electric field, γ is the propagation constant, and ω is the radian frequency. By applying the boundary conditions to the structure in Fig. 2.1,

C. Nguyen and S. Kim, *Theory, Analysis and Design of RF Interferometric Sensors*,
SpringerBriefs in Physics, DOI 10.1007/978-1-4614-2023-1_2,
© Springer Science+Business Media, LLC 2012

Fig. 2.1 Electromagnetic waves traveling in a dielectric characterized by dielectric constant ε_1. ε_0 and ε_2 are the dielectric constants of the preceding and following media, respectively

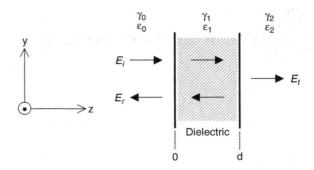

we can obtain neglecting secondary reflections and transmissions at the boundaries [20]:

$$E_0(1 + \Gamma_0) = E_1(1 + \Gamma_1)$$

$$\frac{E_0}{Z_0}(1 - \Gamma_0) = \frac{E_1}{Z_1}(1 - \Gamma_1)$$

$$E_1(e^{\gamma_1 d} + \Gamma_1 e^{-\gamma_1 d}) = T_0 E_0$$

$$\frac{E_1}{Z_1}(e^{\gamma_1 d} - \Gamma_1 e^{-\gamma_1 d}) = T_0 \frac{E_0}{Z_2} \tag{2.2}$$

where E's represent the electric fields in the media; Γ's and T's are the reflection and transmission coefficients, respectively, Z's denote the intrinsic impedances of the media; the subscript number corresponds to each medium; and d is the thickness of the dielectric. By solving (2.2) for Γ_0 and T_0, we can calculate the electric fields E_r and E_t of the reflected and transmitted waves as

$$E_r = \Gamma_0 E_0 \tag{2.3}$$

where

$$\Gamma_0 = \frac{\left(\frac{Z_2}{Z_0} - 1\right)\cosh(\gamma_1 d) - \left(\frac{Z_1}{Z_0} - \frac{Z_2}{Z_1}\right)\sinh(\gamma_1 d)}{\left(\frac{Z_2}{Z_0} + 1\right)\cosh(\gamma_1 d) - \left(\frac{Z_1}{Z_0} + \frac{Z_2}{Z_1}\right)\sinh(\gamma_1 d)}.$$

and

$$E_t = T_0 E_0 \tag{2.4}$$

where

$$T_0 = \left\{ \frac{1}{2}\left[\left(1+\frac{Z_0}{Z_2}\right)\cosh(\gamma_1 d) - \left(\frac{Z_1}{Z_2}+\frac{Z_0}{Z_1}\right)\sinh(\gamma_1 d)\right]\right\}^{-1}.$$

It is the wave defined by the electric field in (2.3) or (2.4) that constructs the measurement wave in a RF interferometry.

2.2 Determination of Relative Dielectric Constant and Thickness

The relative permittivity or dielctric constant (ε_r) as well as relative permiability (μ_r) characterize the material properties. For lossy materials, the relative dielectric constant can be expressed in a complex form of

$$\widehat{\varepsilon}_r = \frac{\widehat{\varepsilon}}{\varepsilon_0} = \varepsilon_r + j\frac{\sigma}{\omega\varepsilon_0} \tag{2.5}$$

where ε_r is the relative dielectric constant, ε_0 is the dielectric constant of free space, σ is the conductivity of the material, and ω is angular frequency. In practice, it is common to introduce the dielectric loss tangent *tanδ* to account for the material loss in the complex dielectric constant as

$$\widehat{\varepsilon}_r = \varepsilon_r(1 + j\tan\delta) \tag{2.6}$$

where the dielectric loss tangent is defined as the ratio between the imaginary and real parts of the complex dielectric constant. The relative dielectric constant of a material located in free space can be determined on the basis of phase or amplitude measurement of either reflected or transmitted waves. In the measurement using the reflection method, the dielectric is typically conductor-backed to increase reflected power. In this case, the intrinsic impedance Z_2 is equal to zero. Then, we can simplify the reflection coefficient in (2.3) as [20]

$$\Gamma_0 = \frac{(\gamma_0 - \gamma_1)\exp(-\gamma_1 d) - (\gamma_0 + \gamma_1)\exp(\gamma_1 d)}{(\gamma_0 + \gamma_1)\exp(-\gamma_1 d) - (\gamma_0 - \gamma_1)\exp(\gamma_1 d)} \tag{2.7}$$

making use of the intrinsic impedance Z_1 in terms of the free-space impedance Z_0

$$Z_1 \simeq \frac{Z_0}{\sqrt{\varepsilon_{r1}}} = \frac{\gamma_0}{\gamma_1}Z_0. \tag{2.8}$$

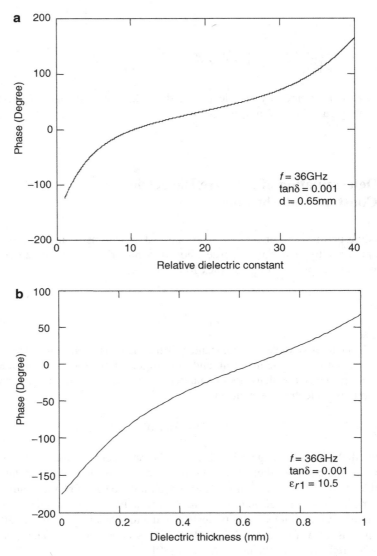

Fig. 2.2 Phase of reflection coefficient versus (**a**) relative dielectric constant and (**b**) dielectric thickness

assuming the dielectric has low loss, where ε_{rl} is the relative dielectric constant of the dielectric to be evaluated. As an example, Fig. 2.2 shows the phase of the reflection coefficient as a function of the relative dielectric constant and thickness. In the measurement using the transmission method, the dielectric is located in free space between two antennas. Therefore, $Z_2 = Z_0$ and $\gamma_2 = \gamma_0$ are satisfied. The transmission coefficient in (2.4) can then be transformed into [20]

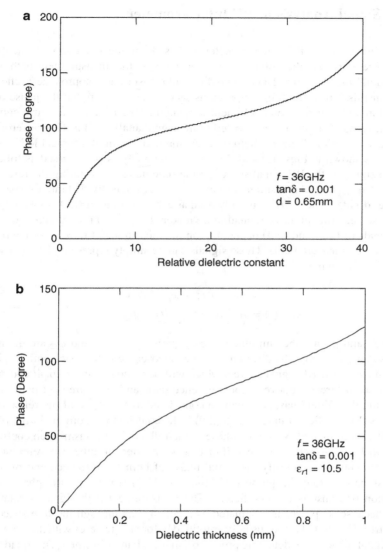

Fig. 2.3 Phase of transmission coefficient versus (**a**) relative dielectric constant and (**b**) dielectric thickness

$$T_0 = \frac{4\gamma_1 \gamma_0}{(\gamma_0 + \gamma_1)^2 \exp(-\gamma_1 d) - (\gamma_0 - \gamma_1)^2 \exp(\gamma_1 d)} \qquad (2.9)$$

Figure 2.3 shows the phase variation of the transmission coefficient given in (2.9) corresponding to the change of the relative dielectric constant and thickness. Note that the reflected and transmitted wave depicted in (2.3) and (2.4), respectively, represent the measurement-path wave in the RF interferometer.

2.3 Signal Analysis of RF Interferometer

The principle of a RF interferometer is based on the detection of the phase difference between the reference-path wave and the measurement-path wave derived in Eqs. (2.3) and (2.4) for two different measurement approaches: reflection and transmission method. In the previous section, it was seen that the phase of the reflection and transmission coefficient is related to the relative dielectric constant and thickness. This section is devoted to the signal analysis of a RF interferometer for phase detection. With the help of a schematic diagram of a typical RF interferometer as shown in Figs. 1.2 and 1.3, the system analysis is discussed as follows.

The signal of the RF signal source in the schematic, constituting the reference-path signal $v_{ref}(t)$ and measurement-path signal $v_{mea}(t)$, is divided into two paths by a power divider. The $v_{ref}(t)$ is usually used as a local oscillator (LO) signal to pump the phase detecting processor, quadrature mixer. The $v_{mea}(t)$ is configured as one of the signals of (2.3) and (2.4) depending on the measurement method (reflection or transmission measurement). Those signals can be simply represented by sinusoidal signals as followings:

$$v_{ref}(t) = A_r \cos(\omega t + \phi_{i1} + \phi_n)$$
$$v_{mea}(t) = A_m \cos[\omega t + \phi(t) + \phi_{i2} + \phi_n] \qquad (2.10)$$

where A_r and A_m are the amplitude of each path signal; ϕ_{i1} and ϕ_{i2} are the initial phases that come from the difference of the electrical length in each path; ϕ_n is the phase noise of the RF signal source, which will be discussed in Chap. 4; and $\phi(t)$ is the phase difference between the reference-path and measurement-path signal, excluding the initial phase, and can be considered as the phase of the reflection or transmission coefficient in (2.7) and (2.9) if the contribution from the initial phases in (2.10) is eliminated. When the phase of the reflection or transmission coefficient needs to be measured, a phase shifter can be inserted in either the reference- or measurement-path to nullify the initial phase of both the reference and measurement signals so that the phase difference in (2.10) reads only the phase of the reflection or transmission coefficient. The measurement-path signal is relatively weak because its power is attenuated as it propagates through the free space and dielectric. This signal is thus usually amplified before it interferes with the reference-path signal in the phase detecting processor. In the RF interferometer, the quadrature mixer is generally used as a phase detecting device. Interfering two different signals, which is performed in the quadrature mixer, can be considered mathematically as a multiplication of these signals. The measurement-path signal, coherently interfered with the reference-path signal and low-pass filtered in the quadrature mixer, produces the following (voltage) output signals in quadrature form:

$$v_I(t) = A_I \cos[\phi(t) + \phi_{i1} + \phi_n]$$
$$v_Q(t) = A_Q \sin[\phi(t) + \phi_{i2} + \phi_n] \qquad (2.11)$$

where subscripts I and Q represent in-phase and quadrature, respectively, and A_I and A_Q are the amplitude of each quadrature signal. By applying inverse trigonometry, we can determine the phase $\phi(t)$, which is the ultimate goal of the interferometric measurement. It is important to notice that the actual response of the quadrature mixer does not exactly follow the form of (2.11) but responds nonlinearly due to its circuit imperfection, which is analyzed in the following chapter. The signals including nonlinearity of the quadratue mixer can be described as

$$v_I(t) = (A + \Delta A) \cdot \cos \phi(t) + V_{OSI}$$
$$v_Q(t) = A \cdot \sin[\phi(t) + \Delta\phi] + V_{OSQ} \tag{2.12}$$

where V_{OSI} and V_{OSQ} are the DC offsets of the I and Q signals, respectively; and ΔA and $\Delta\phi$ represent the amplitude and phase imbalance between the I and Q channels, respectively. For convenience, the initial phase terms and phase noise contribution as seen in (2.10) and (2.11) are ignored here.

From the viewpoint of systems, the function of the quadrature mixer in a RF interferometer is fundamentally homodyne (or direct) down conversion of the measurement-path signal. In addition to the imbalance issues described in (2.12), it is well known that the 1/f noise contribution is a critical problem in the direct down conversion. The best strategy to overcome this problem is to slightly shift the frequency of either the reference-path or measurement-path signal so that the frequency of the mixer's output signal is located far away from the 1/f noise spectrum. The schematic to implement this approach is shown in Fig. 1.4. The two input signals of the phase comparator in Fig. 1.4 are processed independently by two internal quadrature mixers to detect the phase difference between the two signals. Ignoring the initial phase and phase noise effect, we can express the output signals of the quadrature mixer, which is implemented by quadrature sampling digital signal processing technique in our system, as

$$v_I(nT) = A \cdot \cos \phi(nT)$$
$$v_Q(nT) = A \cdot \sin \phi(nT) \tag{2.13}$$

where T is the sampling time of the digital quadrature mixer. The RF interferometer employing this approach is covered in Chap. 4.

Chapter 3
RF Homodyne Interferometric Sensor

Homodyne configuration has been the stereo type of the RF interferometers because of its simplicity. Most RF interferometry, especially for measurement purposes in laboratory, has been developed with this structure. RF interferometry has been used for various applications in instrumentation such as non-destructive characterization of material [20] and plasma diagnostics [21]. RF Interferometery is an attractive means for displacement measurement due to its high measurement accuracy and fast operation. Particularly, it has high resolution due to the fact that the displacement is resolved within a fraction of a wavelength of the operating frequency. Previous works based on optical interferometers have been reported for displacement measurements with resolution ranging from micrometer to sub-nanometer [22–24]. Fast and accurate displacement measurement is needed in various engineering applications such as high-speed metrology, position sensing, liquid-level gauging, water flux sensing, and personal health monitoring.

This chapter presents the development of a millimeter-wave interferometer employing a homodyne configuration and demonstrates its displacement measurement with sub-millimeter resolution as an example of its possible uses for sensing. The system operates at 37.6 GHz and is completely fabricated using microwave and millimeter-wave integrated circuits – both hybrid (MIC) and monolithic (MMIC). The non-linear phase response of the quadrature mixer, which is a critical problem in RF interferometry, is also discussed along with a common I/Q error correction algorithm. The measurement error contributed to the quadrature mixer is estimated by a worst-case error analysis approach.

3.1 System Configuration and Principle

Figure 3.1 shows the overall system block diagram of the homodyne millimeter-wave interferometric sensor. The interferometer transmits a millimeter-wave signal to illuminate a target via the antenna. As depicted in Sect. 2.3, the return signal from the target is captured by the interferometer via the antenna and converted into a

C. Nguyen and S. Kim, *Theory, Analysis and Design of RF Interferometric Sensors*, SpringerBriefs in Physics, DOI 10.1007/978-1-4614-2023-1_3, © Springer Science+Business Media, LLC 2012

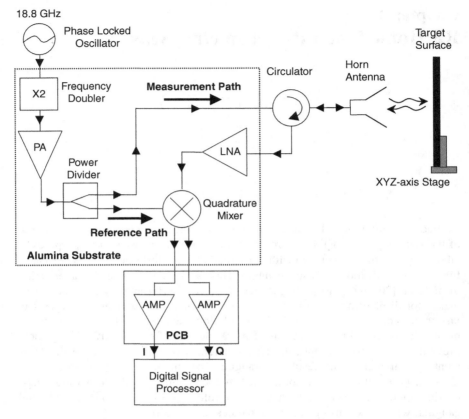

Fig. 3.1 Overall system configuration of the homodyne millimeter-wave interferometric sensor

base-band signal, which is then processed to determine the displacement of the target location.

Displacement measurement using the RF interferometry technique is basically a coherent phase-detection process using a phase detecting processor, which is the quadrature mixer in the system. The phase difference between the reference and measurement paths, produced by a displacement of the target location, is determined from the in-phase (I) and quadrature (Q) output signals of the quadrature mixer. These signals are described as

$$v_I(t) = A_I \sin \phi(t)$$
$$v_Q(t) = A_Q \cos \phi(t) \tag{3.1}$$

where A_I and A_Q are the maximum amplitudes of I and Q signals, respectively. $\phi(t)$ represents the phase difference and can be determined, for an ideal quadrature mixer, as

$$\phi(t) = \tan^{-1}\left(\frac{v_I(t)}{v_Q(t)} \frac{A_Q}{A_I}\right) \tag{3.2}$$

Practical quadrature mixers, however, have a nonlinear phase response due to their phase and amplitude imbalances as well as DC offset. A more realistic form of the phase including the nonlinearity effect can be expressed as

$$\phi(t) = \tan^{-1}\left(\frac{1}{\cos \Delta\phi} \frac{A}{(A + \Delta A)} \frac{v_I(t) - V_{OSI}}{v_Q(t) - V_{OSQ}} - \tan \Delta\phi\right) \tag{3.3}$$

by solving Eq. 2.12 for $\phi(t)$. The detected phase is generated by the time delay, τ, due to round-trip traveling of the electromagnetic wave for the distance between the antenna aperture and target. It has a relationship with range, r, as following:

$$\phi(\tau) = 2\pi f_0 \tau = \frac{4\pi f_0 r}{c} \tag{3.4}$$

where f_0 and c are the operating frequency and speed in free space of the electromagnetic wave. From Eq. 3.4, the range as a function of time variable can be defined by

$$r(t) = \frac{\phi(t)}{4\pi} \lambda_0 \tag{3.5}$$

where λ_0, defined by c/f_0, is the operating wavelength in air, and normal incidence of the wave is assumed. Note that the detected phase corresponds to a round-trip travel of the received signal. Range variation is produced by changes in target location and can be expressed in the time domain as

$$\Delta r(nT) = r[nT] - r[(n-1)T] \quad n = 1, 2, 3, \ldots \tag{3.6}$$

where T is the sampling interval. The displacement for the entire target measurement sequence can be described as a summation of consecutive range variations as

$$d(nT) = \sum_{n=1}^{k} \Delta r(nT) \quad n = 1, 2, 3, \ldots, k \tag{3.7}$$

These range variations can be measured from the data acquisition and processing of the quadrature mixer's output signals, from which an actual displacement can then be constructed. In this displacement construction process, the range ambiguity problem arises due to the 2π-phase discontinuity of the phase detecting processor, which is typically expected in the interferometry technique. This problem is overcome by employing the phase unwrapping signal-processing technique described in [25–27]. Measured data produced by the RF interferometer are wrapped into the range $(-\pi, \pi)$, and the phase unwrapping algorithm is used to reconstruct the wrapped phase beyond the range of $(-\pi, \pi)$ so as to obtain a continuous phase without the 2π radian ambiguities.

3.2 Phase Unwrapping Signal Processing

The phase unwrapping is an essential signal processing technique in interferometric radar. It is applied mainly for synthetic aperture radar (SAR) interferometry, magnetic resonance imaging (MRI) and astronomical imaging. The interferometric signals generated by the phase detecting device, which is the quadrature mixer in the developed system, are wrapped into the range $(-\pi, \pi)$. The goal of phase unwrapping signal processing is to reconstruct the wrapped phase beyond the range of $(-\pi, \pi)$. Mathematically, the phase unwrapping operation is described by the following equation in discrete time domain

$$\Phi(n) = \phi(n) + 2\pi k(n) \tag{3.8}$$

where $\phi(n)$ is an unwrapped phase which is the quantity to be detected, and $k(n)$ is an integer function that enforces $\phi(n)$ wrapped.

Several digital techniques [25–27] have been proposed to develop the phase unwrapping algorithms. Itoh developed a brief and suggestive technique for a one-dimensional case [25]. For a brief discussion of Itoh's method, let us first introduce two operators W and Δ. The operator W wraps the phase into the range $(-\pi, \pi)$

$$W\{\phi(n)\} = \Phi(n) + 2\pi k(n), n = 0, 1, \quad ..., \quad N - 1 \tag{3.9}$$

where $k(n)$ is an integer array selected to satisfy $-\pi < \Phi(n) \leq \pi$. The difference operator Δ is defined as

$$\Delta\{\phi(n)\} = \phi(n + 1) - \phi(n)$$
$$\Delta\{k(n)\} = k(n + 1) - k(n) \quad n = 0, 1, ..., N - 1. \tag{3.10}$$

From the difference of wrapped phase sequences using Eqs. 3.9 and 3.10, we can get

$$\Delta\{W\{\phi(n)\}\} = \Delta\{\phi(n)\} + 2\pi\Delta\{k_1(n)\}. \tag{3.11}$$

Applying the wrapping operation again to the above yields [25]

$$W\{\Delta\{W\{\phi(n)\}\}\} = \Delta\{\phi(n)\} + 2\pi[\Delta\{k_1(n)\} + k_2(n)] \tag{3.12}$$

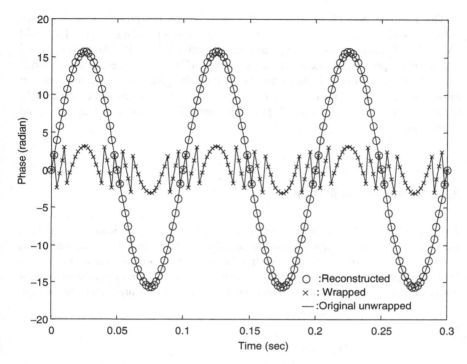

Fig. 3.2 Original unwrapped, wrapped and reconstructed phase sequences

where $k_1(n)$ and $k_2(n)$ distinguish the integer arrays produced by the two consecutive wrapping operations. Equation 3.12 implies that $\Delta\{k_1(n)\} + k_2(n)$ should be zero to satisfy the requirement of $-\pi < \Delta\{\phi(n)\} \leqslant \pi$. Thus it is reduced to

$$W\{\Delta\{\Phi(n)\}\} = \Delta\{\phi(n)\} \tag{3.13}$$

Finally, the integration form of Eq. 3.13 shows that

$$\phi(m) = \phi(0) + \sum_{n=0}^{m-1} W\{\Delta\{W\{\phi(n)\}\}\}. \tag{3.14}$$

Equation 3.14 implies that the actual phase sequences can be unwrapped by iterative integration operation of the wrapped difference of wrapped phases.

Figure 3.2 illustrates an example of phase unwrapping for a one-dimensional case. The sinusoidal phase sequence of Eq. 3.15 with maximum phase variation of 5π is perfectly reconstructed by phase unwrapping operation

$$\phi(t) = 5\pi \sin(2\pi f t). \tag{3.15}$$

The f in the parenthesis implies a periodicity of the phase signal, such as vibration which may come from a periodic displacement of the target. The solid line in Fig. 3.2 represents the original phase function of (3.15). The reconstructed and wrapped phase sequences are designated by (O) and (×), respectively. Typically, the wrapped phase is the form obtained by the phase detecting processor in most interferometric sensor. As shown, the unwrapped phase sequences are exactly reconstructed from the wrapped phase by applying phase unwrapping signal processing.

3.3   System Fabrication

The homodyne millimeter-wave interferometric sensor was fabricated using MICs and MMICs. All components inside the dotted lines shown in Fig. 3.1 are integrated on a 0.254-mm thick alumina substrate using surface-mount technology. The Wilkinson power divider, a counterpart of the beam splitter in optical interferometry as shown in Fig. 1.1, was realized using microstrip lines and analyzed through

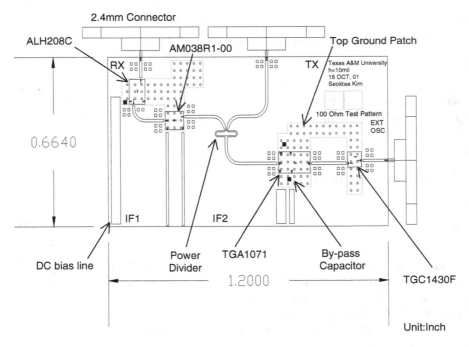

Fig. 3.3 Detailed circuit layout of the homodyne millimeter-wave interferometric sensor. The components corresponding to the part numbers are listed in the text. Dimensions are in inch

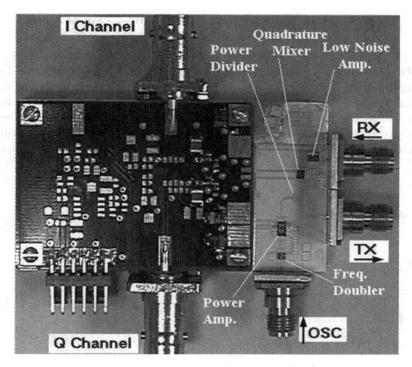

Fig. 3.4 Photograph of the fabricated homodyne millimeter-wave interferometric sensor

field simulation using a commercial field simulator, IE3D [28], and implemented on the top of the alumina substrate employing thin film technology to direct the millimeter wave signal into the reference and measurement paths. The return loss and isolation were optimized to achieve better than 15 and 30 dB, respectively, at frequencies from 36 to 38 GHz. The resistor in the power divider was implemented with Ta_2N thin film and adjusted accurately to obtain its final value, 100 Ω, using laser trimming. Figure 3.3 shows the layout of the millimeter wave circuit in detail. Commercially available Ka-band (26.5–40 GHz) MMICs were used for the quadrature mixer (Alpha industries, AM038R1-00), low noise amplifier (TRW, ALH208C), power amplifier (TriQuint, TGA1071-EPU) and frequency doubler (TriQuint, TGC1430F-EPU); they are surface-mounted on metallic patches, which are gold plated and connected to the alumina's ground plane by 0.2-mm-diameter vias. A Printed Circuit Board (PCB) is used to mount a high-precision operational amplifier, which constitutes a 100-Hz pass-band active low pass filter. The filter provides gain for the output signals of the quadrature mixer and limits the signal bandwidth to reduce the noise-floor level. 3-by-0.5-mil gold ribbons are used to connect the 10-mil-wide alumina transmission lines to the signal pads on the MMICs. Figure 3.4 is a photograph of the fabricated system.

3.4   Displacement Measurement and Liquid-Level Gauging

The developed homodyne millimeter-wave interferometric sensor was used for displacement measurement and liquid-level gauging as examples demonstrating its capability for sensing applications. The measurement was performed using two laboratory test samples. An 18.8-GHz phase-locked source and a Ka-band standard horn antenna were used in the tests. The first sample is a metal plate mounted on a XYZ-axis stage. The XYZ-axis stage has a fine variation precision of 0.01 mm, a high accuracy of 0.002032 mm/25.4 mm and a good repeatability of 0.00127 mm.

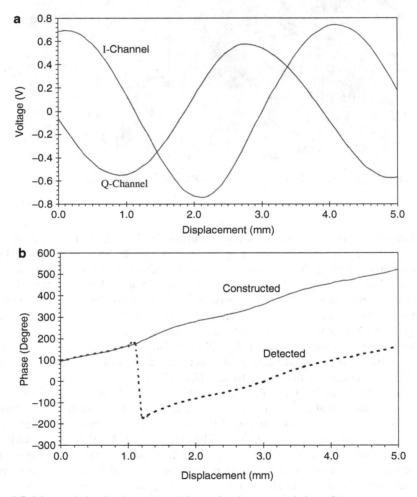

Fig. 3.5 Measured signal voltage (**a**) and detected and constructed phase (**b**)

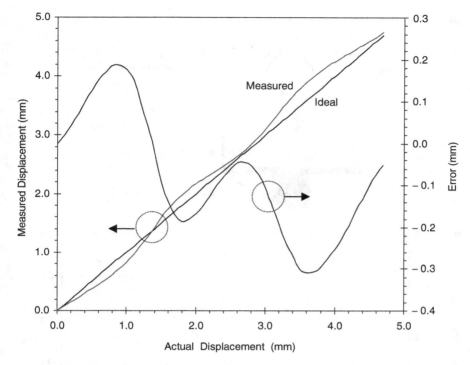

Fig. 3.6 Measured displacement for a metal plate

The metal plate was located 30-cm away from the antenna aperture, and the displacement measurement was made as the plate was moved every 0.1 mm. Signals from the quadrature mixer were captured by the data acquisition hardware (National Instruments, PCI-6111E) with the sample speed of 1 kS/s and sample number of 1,000. Then the entire set of samples is averaged to cancel out the noise components, which are composed of phase noise of the microwave signal source and white noise generated by circuits in the system. Figure 3.5a shows the measured signal voltages, excluding DC-offset voltage, needed for the phase unwrapping. Figure 3.5b displays the phase detected and constructed by the phase unwrapping technique. The phase detected was determined from

$$\phi(t) = \tan^{-1}\left(\frac{v_I(t) - V_{OSI}}{v_Q(t) - V_{OSQ}}\right) \tag{3.16}$$

which contains the errors resulting from the amplitude and phase imbalances of the quadrature mixer. As can be seen in Fig. 3.5b, the reconstructed phase varies from 95° to 523° for a displacement of 5 mm. This range of phase variation is sufficient to validate the phase unwrapping signal processing for phase reconstruction without the 360° ambiguities. For displacements corresponding to multiple times of 360°,

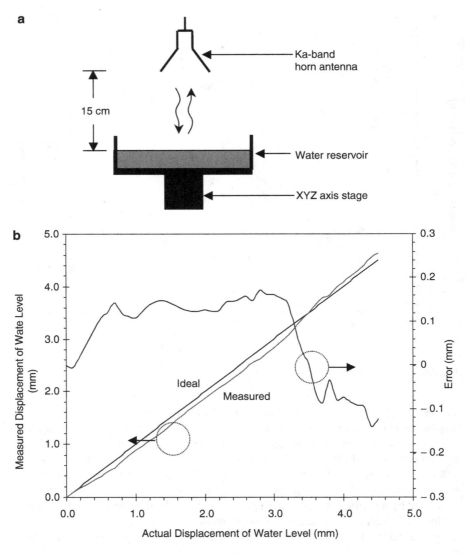

Fig. 3.7 Measurement set-up for water level gauging (**a**) and test results (**b**)

repetition of the phase unwrapping process is needed to construct the phase. The final displacement result is shown in Fig. 3.6 together with the measurement error.

The second sample, as shown in Fig. 3.7a, is water stored in a reservoir, which is mounted on a XYZ-axis stage. It is used to demonstrate a possible application of liquid-level gauging. The water level was located at a distance of 15 cm from the horn antenna and the measurement was made as the distance was varied. Figure 3.7b shows the measured displacement and corresponding error. In both measurements, the homodyne millimeter-wave interferometric sensor achieves a measured resolution of only 0.05 mm and a maximum error of 0.3 mm at each displacement. The resolution was determined through the measurement of the minimum detectable voltage (or phase) as the displacement was varied.

3.5   Analysis of Error Contributed by Quadrature Mixer

3.5.1   Quadrature Mixer Transfer Function

A quadrature mixer is the most common component used to detect phase in RF interferometers based on homodyne structure. Figure 3.8 shows a typical functional block diagram of quadrature mixers. It consists of basically two identically balanced mixers sharing a common in-phase RF input signal and a quadrature phase LO signal to pump the mixers. The LO signal is splitted by a 90° hybrid and the divided signals are fed into the mixers. The conventional problem of a quadrature mixer, as a phase detecting processor, is that it is difficult to achieve good balance for the I and Q paths in terms of amplitude and phase, due to the imperfection of circuit components. This problem is usually called I/Q error, resulting in limitation in accuracy of interferometric measurement. As the operating frequency is increased, the problem becomes severe and hard to control.

Equation 3.17 describes mathematically the ideal response of the quadrature mixer from the I and Q ports, assuming that the frequency of the LO and RF signal is slightly offset by f_{IF}, i.e., $f_{RF} - f_{LO} = f_{IF}$, as

$$v_I(t) = A \cdot \cos(2\pi f_{IF}t + \phi)$$
$$v_Q(t) = A \cdot \sin(2\pi f_{IF}t + \phi)$$

(3.17)

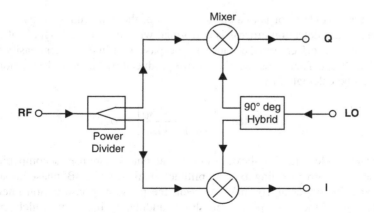

Fig. 3.8 Functional block diagram of quadrature mixers

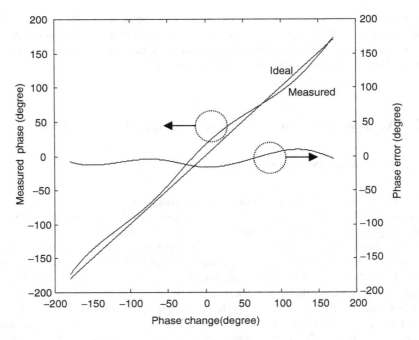

Fig. 3.9 Example of non-linear phase response of a quadrature mixer

where ϕ is the phase information to be detected. Real response, however, is affected by the mixer's imperfection of the amplitude and phase imbalance, causing I/Q error. The time response of the real quadrature mixer can be expressed as

$$v_I(t) = (A \cdot \Delta A) \cos(2\pi f_{IF} t + \phi + \Delta\phi) + V_{OSI}$$
$$v_Q(t) = A \sin(2\pi f_{IF} t + \phi) + V_{OSQ} \tag{3.18}$$

where the amplitude error is treated as the ratio of the amplitudes for two signals. The actual phase detection process is performed with the detected signals of (3.18) excluding the DC offset terms, because a band-pass filter in the system easily filters out the DC offset. Therefore, the phase error produced by the non-ideal quadrature signals can be calculated by

$$\phi_e = \tan^{-1}\left[\frac{v_Q(t) - V_{OSQ}}{v_I(t) - V_{OSI}}\right] - \phi. \tag{3.19}$$

As an example, Fig. 3.9 shows a non-linear phase response, accompanied by the phase error corresponding to an amplitude imbalance of 2 dB, phase imbalance of 10°, and DC offset of 100 mV. As seen, the phase response is non-linear for linear change of the input phase and shows undulating behavior, which implies deterioration of measurement accuracy.

In homodyne systems, the frequency of the RF and LO signals is the same, so that the output of the quadrature mixer generates only DC terms. This makes the estimation of the phase error difficult. It is a common approach to introduce a test signal, which is produced by mixing the RF signal with the LO signal slightly different from the RF signal in frequency so that the output of the mixer is AC of intermediate frequency, f_{IF}. Usually, f_{IF} is chosen low enough to be easily processed by digital signal processing. Based on the test signals, it is possible to estimate the amplitude and phase imbalance as well as the DC offset of the quadrature mixer in frequency domain using Fourier transform, as explained in detail in the following section.

It is convenient to handle the pair of signals in (3.18) as a complex signal; that is,

$$v(t) = v_I(t) + jv_Q(t). \tag{3.20}$$

Applying Fourier transform on (3.20) produces an impulse (delta) function in frequency domain. For an ideal quadrature mixer, the impulse function appears only at the frequency of f_{IF}. But imbalances of the quadrature mixer cause image response at the negative (or image) frequency of f_{IF}. The Fourier transform of complex output signals of a real quadrature mixer is expressed as

$$
\begin{aligned}
\mathrm{F}[v(t)] = {}& \mathrm{F}\big[v_I(t) + jv_Q(t)\big] \\
= {}& \mathrm{F}(0) && : DC\ term \\
& + \frac{1}{2}A \cdot \exp(j\phi)[\Delta A \cos(\Delta\phi) + j\Delta A \sin(\Delta\phi) - 1]\delta(f + f_{IF}) && : Image\ Signal \\
& + \frac{1}{2}A \cdot \exp(j\phi)[\Delta A \cos(\Delta\phi) + j\Delta A \sin(\Delta\phi) + 1]\delta(f - f_{IF}) && : Primary\ Signal
\end{aligned}
\tag{3.21}
$$

which constitutes a DC term that comes from DC offset, upper-side (Primary) and lower-side (Image) signals. The power ratio of the lower- to upper-side signal, namely Image-to-Signal Ratio (*ISR*), measures the amount of deviation of the real quadrature mixer's response compared to an ideal response, and is given by

$$ISR = \frac{[\Delta A \cos(\Delta\phi) - 1]^2 + [\Delta A \sin(\Delta\phi)]^2}{[\Delta A \cos(\Delta\phi) + 1]^2 + [\Delta A \sin(\Delta\phi)]^2}. \tag{3.22}$$

From Eq. 3.22, the amplitude imbalance is derived as [29]

$$\Delta A = \frac{\cos(\Delta\phi)(1 + ISR) + \sqrt{[\cos(\Delta\phi)(1 + ISR)]^2 - (1 - ISR)^2}}{1 - ISR}. \tag{3.23}$$

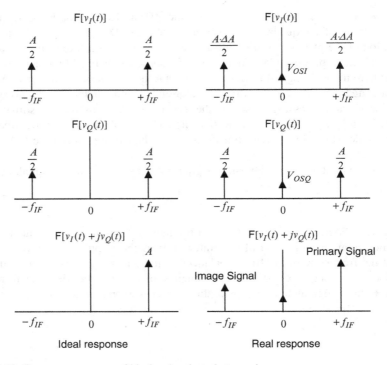

Fig. 3.10 Frequency response of ideal and real quadrature mixer

It is relatively easier to measure *ISR* and *ΔA* using a spectrum analyzer than to measure the phase imbalance *Δφ*. From the measured *ISR* and amplitude imbalance, the phase imbalance can be deduced as [29]

$$\Delta\phi = \cos^{-1}\left[\frac{(\Delta A^2 + 1)(1 - ISR)}{2\Delta A(1 + ISR)}\right]. \tag{3.24}$$

Figure 3.10 illustrates the frequency response of a complex output signal of a quadrature mixer for both ideal and real cases. The image signal shown in the figure is generally called *Hermitian* image; it produces a false target and deteriorates resolution in radar used for most ranging applications. Also, in RF interferometry, it causes a non-linear phase response, as already shown in Fig. 3.9. It is thereby desirable to suppress or eliminate the image signal.

3.5.2   I/Q Error Correction Algorithm

As discussed in the previous section, the image signal influences the quadrature phase detection in RF systems. In most radar and communication applications, it is desirable to correct the I/Q error. In this section, the most common method to correct I/Q error is presented by means of correction coefficients derived from a test signal [30].

We can express the quadrature signals with I/Q errors as

$$v_I(t) = (A + \Delta A) \cdot \cos(2\pi f_{IF} t)$$
$$v_Q(t) = A \cdot \sin(2\pi f_{IF} t + \Delta \phi) \qquad (3.25)$$

where the DC offset is excluded because it can be simply determined by the averaged DC level (or zero frequency component in Fourier transform) of each quadrature channel signal.

The problem of I/Q error correction is analogous to the Gram-Schmidt orthogonalization, designating the quadrature signals with vector matrix notation as

$$\begin{bmatrix} v'_I(t) \\ v'_Q(t) \end{bmatrix} = \begin{bmatrix} S & 0 \\ R & 1 \end{bmatrix} \begin{bmatrix} v_I(t) \\ v_Q(t) \end{bmatrix} \qquad (3.26)$$

where S and R are the rotating and scaling coefficients, respectively, to make the quadrature signals of (3.25) orthogonal; that is, exactly 90° out of phase with equal amplitude. In [30], a digital signal processing technique is suggested to obtain estimates of the coefficient matrix using DFT (Discrete Fourier Transform). By definition, the DFT of complex quadrature signal $v(t)$ yields

$$\mathrm{F}\left(\frac{k}{NT}\right) = \frac{1}{N} \sum_{n=0}^{N-1} v(nT) \exp\left(-j\frac{2\pi kn}{N}\right) \qquad k = 0, 1, 2, \ldots, N-1 \qquad (3.27)$$

where T is the sampling time, N is the number of samples, and $v(nT) = v_I(nT) + jv_Q(nT)$. The complex signal $v(nT)$ can be described from (3.25) as

$$v(nT) = (A + \Delta A) \cdot \cos[2\pi f_{IF}(nT)] + jA \cdot \sin[2\pi f_{IF}(nT) + \Delta \phi]. \qquad (3.28)$$

The primary and image signal components appear at $\mathrm{F}(1/NT)$ and $\mathrm{F}[(N-1)/(NT)]$ and are related to the amplitude and phase imbalance by

$$\mathrm{F}\left(\frac{1}{NT}\right) = \frac{A}{2}[(1 + \Delta A) + \cos(\Delta \phi) + j\sin(\Delta \phi)]$$

$$\mathrm{F}\left(\frac{N-1}{NT}\right) = \frac{A}{2}[(1 + \Delta A) - \cos(\Delta \phi) + j\sin(\Delta \phi)], \qquad (3.29)$$

after substituting (3.28) into (3.27) and solving for the DFT components at the frequencies of $1/NT$ and $(N-1)/NT$. The estimates of the coefficients in (3.26) can then be obtained by the components of DFT of (3.28) as [30]

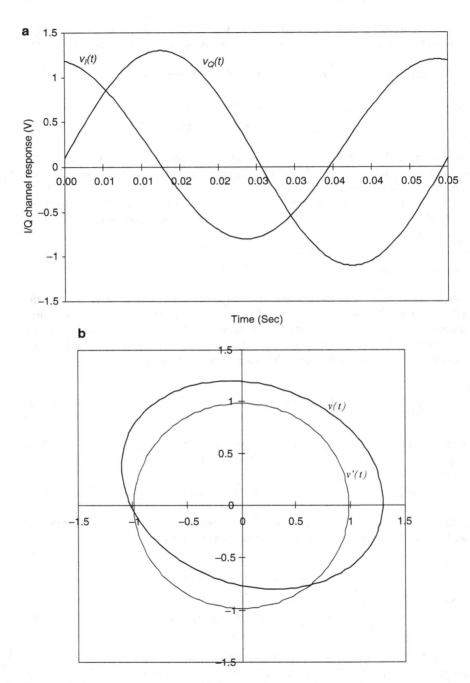

Fig. 3.11 I/Q error correction: (**a**) I/Q channel response and (**b**) geometric interpretation of I/Q error correction

$$\hat{S} = -\mathrm{Re}\left\{\frac{2\mathrm{F}\left[\frac{(N-1)}{NT}\right]}{\mathrm{F}^*\left(\frac{1}{NT}\right) + \mathrm{F}\left[\frac{(N-1)}{NT}\right]}\right\} + 1$$

$$\hat{R} = -\mathrm{Im}\left\{\frac{2\mathrm{F}\left[\frac{(N-1)}{NT}\right]}{\mathrm{F}^*\left(\frac{1}{NT}\right) + \mathrm{F}\left[\frac{(N-1)}{NT}\right]}\right\} \tag{3.30}$$

where F^* means a conjugate of F.

Figure 3.11 shows an example of I/Q error correction by the coefficients obtained by DFT, where the amplitude imbalance is 0.8 dB, phase imbalance is 10°, and DC offset voltage is 100 mV for I channel and 200 mV for Q channel, respectively. Figure 3.11a shows the quadrature signals corresponding to the imbalances mentioned above, and Fig. 3.11b demonstrates geometric interpretation of the correction process. If we plot the in-phase signal $v_I(nT)$ in X-axis and the quadrature signal $v_Q(nT)$ in Y-axis together, then they constitute an ellipse in the XY complex plane. From a geometric viewpoint, the procedure for correcting I/Q error can be interpreted as rotation and scaling of the ellipse in the XY complex plane, as shown in Fig. 3.11b, so that it finally turns into a perfect circle centered at the origin. The coefficients of S and R defined in (3.26) can be geometrically interpreted as the scaling and rotation coefficients to convert the right-most vector $v(t)$ in (3.26), which is neither orthogonal nor equal in amplitude, into the left-most vector $v'(t)$, which is orthogonal and equal in amplitude as in the centered circle in Fig. 3.11b.

3.5.3   Worst-Case Error Analysis

The measurement error employing the homodyne configuration is attributed mostly to the nonlinear response of the quadrature mixer. Several techniques have been proposed to correct the non-linearity of the quadrature mixer [30, 31]. Possible sources of error also come from the measurement distance, the target's reflecting surface, or a combination of these. The instability of the frequency source should produce a negligible effect on the measurement due to the short time delay between the transmit and receive signals of the system. The system can also operate at larger ranges with proper transmitting power.

Measurement accuracy was estimated by an analysis of the maximum phase imbalance using the method proposed in [32]. In this method, the maximum phase error resulted from the image rejection level of the quadrature is calculated. From Eq. 3.22, the image-to-signal ratio (ISR) given as a function of the amplitude and phase imbalance can be reduced to

$$ISR = \frac{1 + \Delta V^2 - 2\Delta V \cos(\Delta\phi)}{1 + \Delta V^2 + 2\Delta V \cos(\Delta\phi)}. \tag{3.31}$$

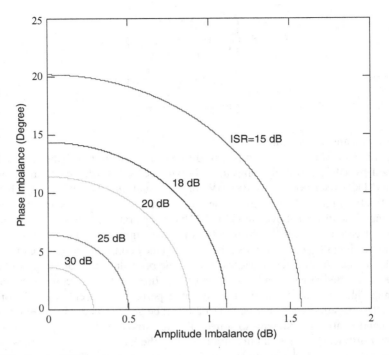

Fig. 3.12 Constant ISR contours

Equation 3.31 can be approximated by an ellipse

$$\left(\frac{\Delta\phi}{X}\right)^2 + \left(\frac{\Delta V}{Y}\right)^2 = 1 \tag{3.32}$$

where

$$X = \cos^{-1}\left(\frac{1 - ISR}{1 + ISR}\right)$$
$$Y = \frac{1 + \sqrt{ISR}}{1 - \sqrt{ISR}}.$$

Figure 3.12 shows constant ISR contours for several different ISR values of a quadrature mixer. As can be seen, the maximum phase imbalance occurs when the amplitude imbalance is 0 dB. Using an ISR of 18 dB from the employed quadrature mixer's data sheet [33], a maximum phase error was obtained as 14.4°, which corresponds to a maximum distance accuracy of 0.32-mm. The measured error of 0.3 mm shown in Figs. 3.6 and 3.7b falls within the maximum calculated error.

Table 3.1 Comparison of the developed sensor's performance with those of commercial liquid-level gauging sensors

	Developed sensor	Saab (tank radar)	Pepperl + Fuchs
Sensing technique	Interferometry	FMCW	FMCW
Operating frequency	37.6 GHz	X-Band (8–12 GHz) Bandwidth: 1 GHz	24 GHz Bandwidth: 0.2 GHz
Accuracy	0.3 mm	1 cm	>20 cm

3.6 Summary

A millimeter-wave interferometric sensor with homodyne configuration, operating 37.6 GHz, was developed. As examples to demonstrate its usage for sensing applications, the sensor was used for accurate displacement sensing and liquid-level gauging. The sensor was integrated on alumina substrate and PCB, employing MMICs and MICs. It has been used to measure accurately the displacement of metal plate location and water level. From these measurement results, it has been found that sub-millimeter resolution in the order of 0.05 mm is feasible. A measurement accuracy of 0.3 mm was also obtained and is within the maximum error calculated on the basis of worst-case error analysis for the I/Q error of the quadrature mixer. The developed sensor's performance is compared with those of some commercial liquid-level gauging sensors in Table 3.1, where the specifications of the commercial sensors are referred to in [34]. The measured results demonstrate the workability of the developed sensor and its potential as an effective tool not only for displacement measurement and liquid level-gauging but also for other sensing applications.

Chapter 4
Double-Channel Homodyne Interferometric Sensor

The previous chapter was devoted to the millimeter-wave interferometric sensor with homodyne configuration. This sensor was used for displacement measurement and liquid level gauging and achieved a resolution of 50 μm, which is equal to $\lambda_0/160$, with λ_0 being the free-space operating wavelength, and 0.3-mm maximum error.

In this chapter, we present a millimeter wave interferometer using double-channel homodyne configuration. This sensor configuration can eliminate the non-linear phase response of the quadrature mixer, which critically limits the sensor's measurement accuracy. The prominent difference between the double-channel homodyne and homodyne configurations is that either measurement- or reference-path signal is modulated using a quadrature up-converter so that the phase information can be detected at an intermediate frequency (IF), which is a frequency low enough to be handled with a digital signal processor. The phase is then detected by the quadrature sampling digital signal processing technique. With this approach, it is possible to exclude the conventional imbalance problem of the quadrature mixer. Also the double-channel homodyne configuration provides additional advantage of estimating the phase noise effect of a RF signal source using FFT algorithm, without the help of phase noise measurement equipment, due to the fact that the phase noise of a RF source is down-converted and appears at IF.

A millimeter-wave interferometric sensor operating at 35.6 GHz was developed based on the double-channel homodyne configuration shown in Fig. 1.4 and demonstrated for possible sensing applications through measurement of displacement and low velocity. Microwave and millimeter-wave Doppler radar has drawn much attention in the automobile industry as a speed-detection sensor for intelligent cruise control, collision-avoidance, and antilock brake systems for vehicles [35–39]. A RF interferometer can be configured to perform the functions of both displacement sensing and velocity measurement, effectively working together as the interferometric displacement sensor and Doppler velocity sensor. The displacement sensing is achieved by configuring the sensor as an interferometric device. The velocity measurement is realized by detection and estimation of the Doppler frequency shift in base band, which is processed against a phase detected by the

C. Nguyen and S. Kim, *Theory, Analysis and Design of RF Interferometric Sensors*,
SpringerBriefs in Physics, DOI 10.1007/978-1-4614-2023-1_4,
© Springer Science+Business Media, LLC 2012

interferometric function of the sensor. In [40], a six-port wave-correlator was developed to achieve the same purpose. The double-channel homodyne millimeter-wave interferometer was realized using MICs and MMICs. Measured displacement results show a resolution of only 10 μm, which is approximately equivalent to $\lambda_0/840$ in terms of free-space wavelength λ_0, a remarkable resolution in terms wavelength. A maximum error of only 27 μm was obtained after corrections using a polynomial curve fitting. Results indicate that multiple reflections dominate the displacement measurement error. For low-velocity measurement, experiments were performed in a laboratory for a moving target on a commercial conveyor. The sensor was able to measure speed as low as 27.7 mm/s, corresponding to 6.6 Hz in Doppler frequency, with an estimated velocity resolution of 2.7 mm/s. A digital quadrature mixer (DQM) was configured as a phase detecting processor, employing the quadrature sampling signal processing technique, to overcome the non-linear phase response problem of a conventional analog quadrature mixer. The DQM also enables low Doppler frequency to be measured with high resolution. The Doppler frequency was determined by applying linear regression on the phase sampled within only fractions of the period of the Doppler frequency. Short-term stability of a RF signal source was also considered to predict its effect on measurement accuracy.

In Doppler velocity measurement, a common method to estimate the Doppler frequency is the maximum likelihood estimate (MLE) obtained by determining the spectral peak centroid in a periodogram, which is implemented by combining Fast Fourier Transform (FFT) algorithm and numerical technique. In the presented double-channel homodyne millimeter-wave interferometer, we employed a different approach using signal processing, based on quadrature phase detection in base band, to estimate the Doppler frequency by applying linear regression on the detected phase. This represents an effective way, particularly for estimating the low-frequency sinusoidal signal needed for low-velocity measurement, compared to the FFT-based MLE. The developed sensor when used for low-velocity measurement has potential to replace the laser Doppler velocimeter, especially in a humid and dusty environment, due to the fact that it is less sensitive than the laser-based velocity sensor to dust particles and water in the air.

4.1 System Configuration and Principle

The overall system configuration is shown in Fig. 4.1. The system is divided into three parts: a millimeter-wave (MMW) subsystem for processing a millimeter-wave signal, an intermediate-signal subsystem for processing signals at intermediate frequencies, and a digital signal processor. The 17.8-GHz phase-locked oscillator, the Ka-band directional coupler, and the lens horn antenna are external components. The sensor transmits a millimeter-wave signal at 35.6 GHz toward a target. The directional coupler, providing good isolation between the transmit and receive ports, is used to direct the signal to the antenna. The signal reflected from

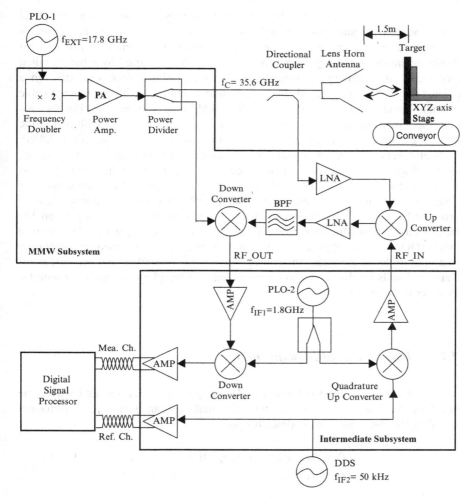

Fig. 4.1 Overall system block diagram of the double-channel heterodyne millimeter-wave inter-ferometric sensor. The target sits either on the XYZ axis (for displacement sensing) or on the conveyor (for velocity measurement). The Reference Channel is not needed for velocity measurement

the target is captured via the antenna, redirected by the coupler to the receiver circuitry, and up-converted by mixing with the RF signal produced by modulating the signal at the first intermediate frequency (IF), f_{IF1}, with the signal of the second intermediate frequency, f_{IF2}, in a direct quadrature up-converter. The up-converted signal is then passed through a coupled-line band pass filter to reject its image component. This signal is combined with part of the transmitted millimeter-wave signal to generate a down-converted RF signal, which is further down-converted by another down-converter in the intermediate-signal subsystem. f_{IF1} and f_{IF2}

are chosen as 1.8 GHz and 50 KHz, respectively. Consequently, the final down-converted signal, $v_M(t)$, namely the measurement-channel signal, contains information on the phase or phase change over time generated by the target displacement or movement, respectively. It is finally amplified by a band-limited differential amplifier and transferred to the digital signal processor, through a twisted cable. The differential driving amplifier combined with the twisted cable provides good noise suppression as well as additional voltage gain. For the displacement measurement, the measured phase of $v_M(t)$ is compared with that of the reference-channel signal, $v_R(t)$, coming from the direct digital synthesizer (DDS). $v_R(t)$ also serves as an IF signal for the direct quadruture up-converter in the intermediate-signal subsystem. If the target is in motion, the frequency of $v_M(t)$ is shifted by the Doppler frequency. In velocity measurement, the phase change over time is detected in the signal processing, and the only measurement-channel signal is processed to extract the Doppler frequency shift. Instead of employing an analog millimeter-wave quadrature mixer as in the previous work [18], a digital quadrature mixer (DQM) was configured as a phase detecting processor based on quadrature sampling to detect the phase difference between the reference- and measurement-channel signals for displacement measurement and the phase change over time for Doppler velocimetry.

4.1.1   Displacement Measurement

Displacement of a target is measured by detecting the phase difference between the two base band signals: reference-channel signal $v_R(t)$ and measurement-channel signal $v_M(t)$. These signals are described as

$$
\begin{aligned}
v_R(t) &= A_R \sin[2\pi f_{IF2}t + \phi_R(t)] + n(t) \\
v_M(t) &= A_M \cos[2\pi f_{IF2}t + \phi_M(t) + \phi_n(t)] + n(t)
\end{aligned}
\tag{4.1}
$$

where A_R, A_M and $\phi_R(t)$, $\phi_M(t)$ are the peak amplitudes and phases of these signals, respectively; $\phi_n(t)$ is the phase noise down-converted from the millimeter-wave signal; and $n(t)$ is the white Gaussian noise. The phase noise of the reference-channel signal is not considered here because its contribution is negligible as compared to that of the measurement-channel signal.

The phase of each channel's signal is obtained by the quadrature sampling signal processing and discussed in the following section. The phase difference between the channels, $\phi_D(t)$, is defined as

$$
\phi_D(t) = \phi_M(t) - \phi_R(t) + \phi_n(t)
\tag{4.2}
$$

Here, we consider only the phase noise, neglecting the white Gaussian noise, on the basis of the phase noise spectrum of the actual signals shown in Fig. 5.4 and obtained by the FFT spectral estimator[1] which indicates that phase noise is the dominant noise source in the frequency band of interest and is approximately greater than 30 dB from the noise floor.

The differential phase difference, needed for calculating the displacement, is obtained in the (digital) time domain as

$$\Delta\phi_D(nT) = \phi_D(nT) - \phi_D[(n-1)T] \quad n = 1, 2, 3, \ldots \tag{4.3}$$

where T is a sampling time interval.

In the case of normal incidence of a wave, the range $r(t)$ from the antenna to the target is related to the phase detected $\phi(t)$ as

$$r(t) = \frac{\phi(t)}{4\pi}\lambda_0. \tag{4.4}$$

The displacement is given by

$$\Delta r(nT) = r[nT] - r[(n-1)T] \quad n = 1, 2, 3, \ldots \tag{4.5}$$

which can be determined using (4.3) and (4.4). The total displacement of the entire target movement is a summation of consecutive displacements:

$$d(nT) = \sum_{n=1}^{k} \Delta r(nT) \quad n = 1, 2, 3, \ldots, k. \tag{4.6}$$

4.1.2 Doppler Velocimetry

The measurement-channel signal, $v_M(t)$, produced by the target in motion, is frequency-shifted in base band by the Doppler frequency, f_d, and can be expressed as

$$v_M(t) = A_M \cos[2\pi f_{IF2}t + 2\pi f_d t + \phi_n(t) + \phi_i] \tag{4.7}$$

where ϕ_i represents the deterministic phase constant.

The principle of radar velocimetry relies on the detection and estimation of the Doppler frequency generated by a moving target. For the normal incident wave, which is our interest, the Doppler frequency is related to the target speed, v, and the wave length, λ, as

$$f_d = \frac{2v}{\lambda} \tag{4.8}$$

---

[1] FFT spectral estimator is a signal processing based on FFT for generating the frequency spectrum of a signal.

in which the target velocity is linearly proportional to the Doppler frequency. The Doppler frequency shift is obtained in base band with reference to the intermediate frequency, f_{IF2}, by taking gradient for time derivative of the detected phase over time.

4.2 Signal Processing

The sensor's signal processing consists of two distinct parts: one for detecting the phase difference needed for measuring the displacement, and another one for estimating the Doppler frequency used for calculating the velocity.

4.2.1 Phase-Difference Detection for Displacement Measurement

Figure 4.2 shows the signal processing flow to extract the phase difference between the measurement- and reference-channel signals in the digital signal processor. In the input signals, the subscripts $+$ and $-$ designate different polarities of the differential signals coming from the differential amplifier in the IF subsystem. The front-end differential amplifier installed in the data acquisition hardware not only amplifies both input channel signals driven by the sensor, but also greatly suppresses the common mode noise with a more than 50-dB common mode rejection ratio, due to the inherent characteristic of a differential amplifier. The reference- and measurement-channel signals are converted into a digital form with 12-bit resolution by the analog-to-digital converter (ADC), implemented in the data acquisition hardware. These signals are expressed as

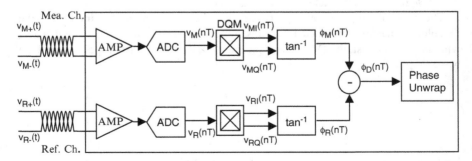

Fig. 4.2 Signal processing flow in the digital signal processor for displacement measurement

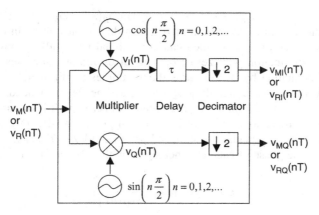

Fig. 4.3 Configuration of the digital quadrature mixer

$$v_R(nT) = A_R \sin[2\pi f_{IF2}(nT) + \phi_R(nT)]$$
$$v_M(nT) = A_M \cos[2\pi f_{IF2}(nT) + \phi_M(nT) + \phi_n(nT)] \qquad n = 1, 2, 3, \ldots \qquad (4.9)$$

A digital quadrature mixer (DQM), based on quadrature sampling signal processing technique was configured and implemented as shown in Fig. 4.3. Various quadrature sampling schemes have been proposed for coherent detection in radar and communication receivers [41–45]. The advantage of the quadrature sampling is that it can eliminate or, at least, minimize the non-linear phase response of a conventional analog quadrature mixer, which is caused by the phase and amplitude imbalances as well as the DC offset voltage of the mixer itself. As the operating frequency is increased, the non-linearity becomes severe and difficult to control. Several correction techniques have also been developed in [30–32]. The DQM implemented in the developed double-channel heterodyne interferometer was inspired by the work presented in [42, 44] and realized by software. The DQM processes each digitized channel signal to generate the in-phase and quadrature components of $v_{MI}(nT)$, $v_{MQ}(nT)$ and $v_{RI}(nT)$, $v_{RQ}(nT)$. The sampling frequency is set as four times the second intermediate frequency, $4f_{IF2}$, so that the digital local oscillators become a quadrature sequence of only -1, 0, and 1, which implies that local oscillators feed exactly 90° out of phase and equal amplitude signals into the mixers, because their phases are integer multiple of $\pi/2$. The mixer designated in Fig. 4.3 performs as a multiplier. The multiplication process samples the following in-phase and quadrature components of the reference-channel signal [46]:

$$v_I(nT) = \begin{bmatrix} 0 \\ I(nT) \cos \phi_i + Q(nT) \sin \phi_i \end{bmatrix} \qquad \begin{matrix} n = odd \\ n = even \end{matrix}$$

$$v_Q(nT) = \begin{bmatrix} 0 \\ Q(nT) \cos \phi_i - I(nT) \sin \phi_i \end{bmatrix} \qquad \begin{matrix} n = even \\ n = odd \end{matrix} \qquad (4.10)$$

where $I(nT) = A_R \cos \phi_R(nT)$, $Q(nT) = A_R \sin \phi_R(nT)$. ϕ_i is the initial phase, which is static in nature. As seen in (4.10), the odd samples of the in-phase signal $v_I(nT)$ and the even samples of the quadrature signal $v_I(nT)$ always produce zero, caused by multiplication with zero from digital oscillators, and they need to be discarded. Decimating by two discards those samples to eliminate zero output in (4.11). In this quadrature sampling approach, a time delay in quadrature signal occurs because the first sample of $v_I(nT)$ produces zero and it is discarded. Therefore, adding a time delay of $\tau = 1/4f_{IF2}$ to the in-phase signal eliminates the time delay between the two quadrature signals, $V_{RI}(nT)$ and $V_{RQ}(nT)$. Taking arctangent then produces the phase of each channel signal within 2π radians, $[-\pi, \pi]$, as

$$\phi_R(nT) = \tan^{-1}\left[\frac{v_{RI}(nT)}{v_{RQ}(nT)}\right]. \tag{4.11}$$

In this configuration, a low-pass filter is not needed for the rejection of the harmonics as in a typical mixer configuration, thus avoiding the filter's transient response to appear in the quadrature outputs, another advantage of our DQM approach.

For the measurement-channel signal, the same procedure is used to obtain

$$\phi_M(nT) = \tan^{-1}\left[\frac{v_{MI}(nT)}{v_{MQ}(nT)}\right]. \tag{4.12}$$

The phase difference, to be converted into displacement, is then determined as

$$\phi_D(nT) = \phi_M(nT) - \phi_R(nT) + \phi_n(nT). \tag{4.13}$$

Finally, the phase-unwrapping process [25–27], explained in Eq. 4.2, is applied to (4.13) to overcome the 2π-discontinuity problem of the phase detection processor. The range corresponding to the detected phase difference is then determined by (4.4) and the displacement is obtained by Eqs. 4.5 and 4.6.

4.2.2   Doppler-Frequency Estimation for Velocity Measurement

Figure 4.4 depicts the signal processing flow used for estimating the Doppler frequency. The measurement-channel signal produced by a target in motion can be expressed, in digital form, as

$$v_M(nT) = A_M \cos\left[2\pi f_{IF2}(nT) + 2\pi f_d(nT) + \phi_n(nT) + \phi_i\right]$$
$$n = 0, 1, 2, \dots, N-1. \tag{4.14}$$

A quadrature down-conversion by the DQM, combined with the phase-based frequency estimation shown in Fig. 4.4, allows low Doppler frequency to be

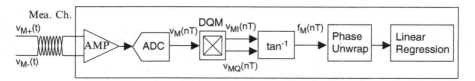

Fig. 4.4 Signal processing flow for velocity measurement

measured with high resolution and directional information, regardless of the number of cycles of the Doppler frequency. A time-varying phase sequence, $\phi_M(nT)$, is generated from the down-converted quadrature signals, $v_{MI}(nT)$ and $v_{MQ}(nT)$. Taking arctangent gives the phase sequence of the down-converted measurement-channel signal within 2π radians, $[-\pi,\pi]$, as

$$\phi_M(nT) = 2\pi f_d(nT) + \phi_n(nT) + \phi_i. \tag{4.15}$$

The phase unwrapping process is then applied to (4.15) to overcome the 2π-discontinuity problem of the phase detection processor. For velocity measurement, the Doppler frequency shift is estimated by applying the least squares or linear regression [47] over the unwrapped phase sequence of (4.15), from which target velocity can be calculated. This approach is, in principle, motivated by the work of Tretter [48]. The process of linear regression fits the unwrapped phase sequence, corrupted by phase noise, into a straight line, from which the Doppler frequency is obtained by taking gradient of the regression line. The phase locking of a RF signal suppresses the 1/f noise component of a reference oscillator, typically a YIG oscillator, and allows the reference oscillator to follow the frequency characteristics of an internal or external frequency standard, usually a temperature-stabilized crystal oscillator, within the phase-locking frequency range. Thus the phase noise spectrum of a phase-locked microwave signal source shows a white Gaussian noise spectrum within the phase-locked bandwidth. In the data acquisition, the sampled data is affected mainly by the white noise down-converted from a RF signal. Based on this fact, the problem of Doppler frequency estimation is transformed into the minimization of the square error [48], by fitting a linear line to $\phi_M(nT)$ corrupted with white Gaussian noise,

$$\varepsilon = \sum_{-(N-1)/2}^{(N-1)/2} \left\{ \phi_M(nT) - \left[2\pi \hat{f}_d(nT) + \hat{\phi}_i \right] \right\}^2 \tag{4.16}$$

where $\hat{f}_d$ and $\hat{\phi}_i$ are the estimates of the Doppler frequency and phase constant, respectively, and N is a total sample number. Random process (noise) is generally treated with statistical analysis tools of mean, standard deviation, and variance. The corresponding theoretical lower limit of variance, Cramer-Rao Bound (CRB), for the frequency estimate $\hat{f}_d$ is derived in [48] as

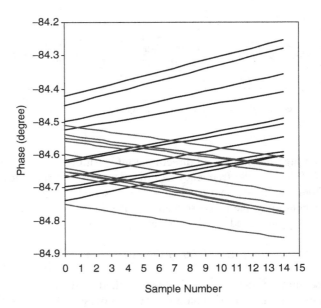

Fig. 4.5 Linear regression for the Doppler frequency of ± 1 Hz

$$CRB(\hat{f}) = \frac{1}{2\pi} \frac{6}{SNR \cdot T^2 N(N^2 - 1)} \qquad (4.17)$$

where high signal to noise ratio (SNR) is assumed, and phase noise is presumed as white. If the error in (4.16) is unbiased, which is valid for high signal-to-noise ratio, then the true Doppler frequency shift can be obtained as

$$f_d = E[\hat{f}_d] \qquad (4.18)$$

where E denotes a statistical expectation or mean.

As an example, Fig. 4.5 illustrates the linear regression performed 10 times for the Doppler frequency of ± 1 Hz, generated by DDS, and $N = 32$. The Doppler frequency is estimated from the gradient of each regression line from phase-time sequences, and the sign of the gradient determines the opening (receding) or closing (approaching) motion of a target.

The corresponding time response of DQM is shown in Fig. 4.6, which was deliberately acquired over many samples ($N = 250,000$) to cover one period of the Doppler frequency of $+1$ Hz. In the developed double-channel heterodyne millimeter-wave interferometric sensor, however, the number of samples used for the linear regression frequency estimator is only a small fraction of that used for one period of the Doppler frequency. On the contrary, relatively large samples in FFT algorithm are required to detect a low-frequency sinusoid with high resolution as seen by the following relationship:

$$\Delta f = \frac{f_s}{N} \qquad (4.19)$$

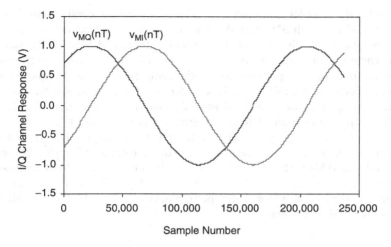

Fig. 4.6 Time response of the DQM for the Doppler frequency of +1 Hz

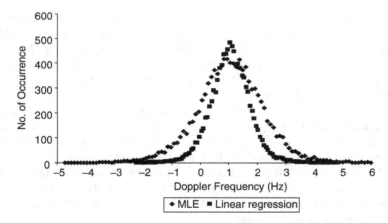

Fig. 4.7 Histogram of the estimated Doppler frequency using MLE and linear regression methods

where Δf and f_S are the resolution and sampling frequency of FFT, respectively. In practice, it cooperates with maximum likelihood estimation (MLE) to maximize the resolution in FFT, which is composed of two steps: first, coarse spectral peak is determined by FFT; and second, fine peak is analyzed by introducing numerical technique like a center of gravity algorithm.

Comparison of the capability between the two different frequency estimators, FFT-based MLE and linear regression, is given in Fig. 4.7, which displays the histogram of the Doppler frequency estimates, iterated 10,000 times for the test signal generated by DDS, and shows the difference in statistical distribution of the estimate for the Doppler frequency of +1 Hz, with the same condition of $f_s = 200$ kHz, $N = 32$ and high SNR (70 dB). The variance of the frequency estimator is dependent on the number of samples, sampling time, and SNR of the sampled signal. Therefore, different conditions imposed on one of those parameters result in differences in estimation performance. In the comparison, the same conditions are exerted and only the high SNR case is considered because, in the sensor, it is easily realizable at f_{IF2} by cascading band-limited amplifiers without much increase in cost. The criterion of high SNR was referred to as 15 dB in [48]. As the figure indicates, the linear regression (on detected phase) shows a narrower statistical distribution, which implies a smaller variance of the estimated Doppler frequency. In the FFT-based MLE, the Center of Gravity algorithm was used for determination of the spectral centroid [49]. As can be seen in Fig. 4.7, the linear regression frequency estimator provides better performance than the FFT-based MLE as long as high SNR is maintained.

4.3   System Fabrication and Test

As seen in Fig. 4.1 and discussed previously, the double-channel heterodyne millimeter-wave interferometric sensor is divided into three parts. The millimeter-wave and intermediate-signal subsystems were realized using MICs and MMICs. The millimeter-wave subsystem was fabricated on a 0.25-mm-thick alumina substrate, as shown in Fig. 4.8a. The intermediate-signal subsystem was implemented on a FR-4 Printed Circuit Board (PCB), as shown in Fig. 4.8b. In the millimeter-wave subsystem, a Wilkinson power divider was designed to split the millimeter-wave signal into the transmit signal and the local oscillator (LO) signal for the down-converter. The band-pass filter is a coupled-line filter. It was designed for a 3-dB bandwidth of about 2 GHz at center frequency 36 GHz using the field simulator IE3D and acts as an image-rejection filter. Details of the alumina circuit layout are shown in Fig. 4.9, where the metallization of a microstrip transmission line is composed of TiW (250 Å), Ni (0.001016 mm), and Au (0.00381 mm) metal combination.

In the millimeter-wave subsystem, commercially available Ka-band MMICs were used for the up-converter (Velocium, MSH108C), down-converter (Velocium, MDB162C), low noise amplifier (Velocium, ALH208C), power amplifier (TriQuint, TGA1071-EPU), and frequency doubler (TriQuint, TGC1430F-EPU). They were surface-mounted on metallic patches connected to the alumina substrate's ground plane by 0.2-mm-diameter vias. These chips were bonded to 0.25-mm-wide microstrip lines using gold ribbons.

In the intermediate-signal subsystem, a phase-locked oscillator operating at 1.8 GHz, designated by PLO-2 in Figs. 4.1 and 4.9b, was designed using a phase-locked-loop frequency synthesizer (Analog Devices, ADF4113) which requires only a low-pass-loop filter as a external component, along with a voltage-controlled oscillator (Sirenza Microdevices, VCO190-1850T) and a 10-MHz oven-controlled

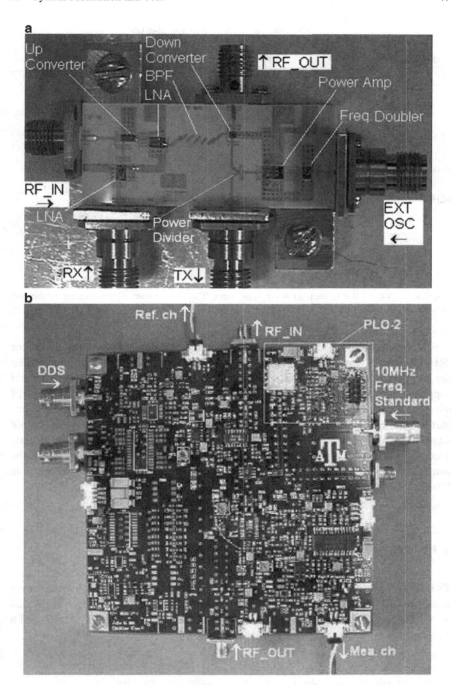

Fig. 4.8 Photograph of the fabricated millimeter-wave (**a**) and intermediate-signal (**b**) subsystems

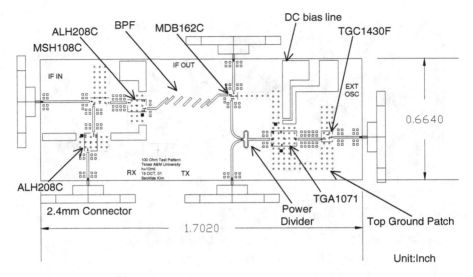

Fig. 4.9 Layout of the millimeter wave subsystem. The components corresponding to the part numbers are listed in the text. Dimensions are in inch

crystal oscillator used as a frequency standard. A direct quadrature modulator (RFMD, RF2422) was used to generate a single sideband (SSB) signal that shifted the frequency f_{IF1} by f_{IF2}. The measured SSB signal shows carrier and sideband suppression of greater than 45 dB at the IF of 50 kHz, achieved by tuning the phase of the IF quadrature input signal. For the down-conversion, a direct quadrature demodulator (Analog Devices, AD8347) was utilized, and one of the quadrature output signals was served as the measurement-channel signal. The differential amplifiers, used for the measurement- and reference-channel signals, greatly suppress the common-mode noise, resulting in a common-mode rejection ratio[2] of more than 50 dB.

4.3.1 Displacement Measurement Results

The developed double-channel heterodyne millimeter-wave interferometric sensor was tested for measuring the displacement of a metal plate mounted on a XYZ axis stage. The stage has a precision of 10 μm, an accuracy of 2.0 μm/25.4 mm, and a repeatability of 1.27 μm.

---

[2] Common-mode rejection ratio is a measure of differential amplifier's ability to reject an undesired signal (noise) that is common to both inverting and non-inverting 180° out-of-phase input terminals.

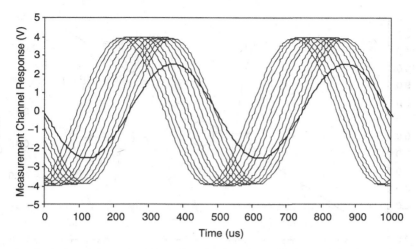

Fig. 4.10 Measurement-channel response for every 100-μm displacement. Thick solid line represents reference-channel signal

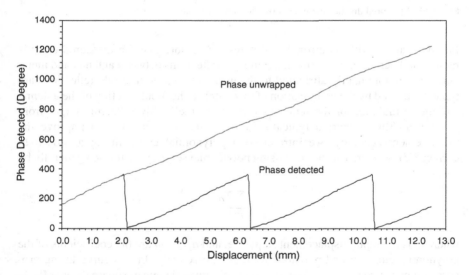

Fig. 4.11 Detected and unwrapped phase

The metal plate was located at 1.5 m away from the antenna aperture. The displacement was measured as the plate was moved every 100 μm. For each measurement, the data acquisition board sampled 1,000 data points and averaged them to cancel out white noise components. Figure 4.10 shows the voltage response of the measurement-channel signal triggered with the reference-channel signal for every 100-μm displacement. The detected and unwrapped phases are shown in Fig. 4.11. The constructed displacement from the unwrapped phase is shown in

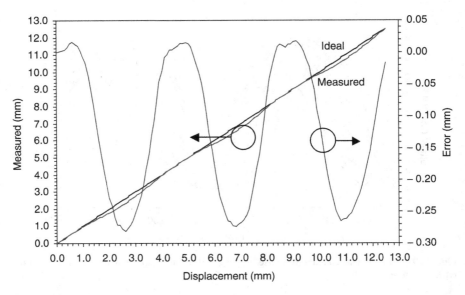

Fig. 4.12 Measured displacement and error for a metal plate

Fig. 4.12 along with its error. It is interesting to note that the measured error is regular and periodic, indicating that multiple reflections between antenna and metal plate are predominantly attributed to the error source; the multiple reflections are typically caused by the combination of mismatch at the input junction of the antenna and the re-radiation of the reflected wave from the highly reflecting target from antenna aperture occurred typically in short distance. In order to improve the measurement accuracy, we introduced a polynomial curve fitting approach to correct for the error. The curve fit using polynomial series, y_i, is formed generally by

$$y_i = \sum_{j=0}^{m} a_j d_i \qquad (4.20)$$

where d_i is the input displacement sequence, and a_j and m are the coefficients of the polynomial curve fit and polynomial order, respectively. In this curve-fitting process, the coefficients a_j are determined to minimize the mean square error (*MSE*),

$$MSE = \frac{1}{N} \sum_{i=0}^{N-1} (y_i - r_i)^2 \qquad (4.21)$$

where r_i represents the measured input sequence, and N is number of the data points.

Figure 4.13 shows the measured and curve-fitted errors with a polynomial order of 13 using SVD (Singular Value Decomposition) algorithm [50]. The displacement

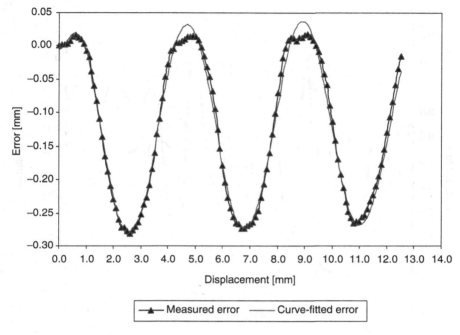

Fig. 4.13 Error correction by the polynomial curve fitting

after correction along with the error is shown in Fig. 4.14. A maximum error of
27 μm was obtained after the correction was made, a significant improvement from a
maximum error of 281 μm without the correction as displayed in Fig. 4.12.
Figure 4.15 shows another measurement result demonstrating the achieved resolution. As shown, the measured maximum error distributes within 10 μm for the entire
displacement of 300 μm. The result indicates that the smallest distance to discrimi-
nate two different positions of a target (i.e., resolution) is only 10 μm, which is
equivalent to about $\lambda_0/840$. It should be mentioned here that in the last measurement,
the XYZ axis stage was moved every 10 μm, which is the precision limit of the stage.
Thus, it is expected that part of the error in accuracy was due to the actual
displacement of the stage, and a motorized stage would provide better reference
for ideal displacement, leading to better accuracy.

4.3.2   Velocity Measurement Result

The velocity of a closing target of metal plate, placed 1.5 m away from the sensor's
antenna aperture, was measured by varying the speed of the conveyor carrying the
plate. The experimental results, shown in Fig. 4.16, were taken consecutively five
times (represented by the measurement index), with N taken as 128. Each time, the

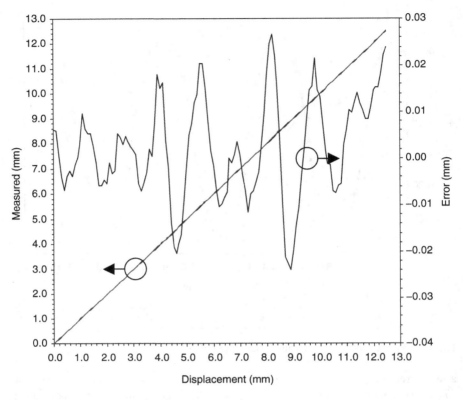

Fig. 4.14 Displacement results after error correction

measurement was repeated 200 times and the results were averaged. The variance of the Doppler frequency estimate for the averaged samples is then reduced by

$$\sigma_{avg}^2(\hat{f}_d) = \frac{\sigma^2(\hat{f}_d)}{N_{avg}} \tag{4.22}$$

where $\sigma^2(\hat{f}_d)$ is the variance of estimate for samples with $N = 128$, and N_{avg} is the number of average; the standard deviation (σ) is a measure to characterize a random process with Gaussian distribution along with mean and variance; the number of the Doppler frequency estimate, which falls within $\pm 1\sigma$, is 68% of the total number of estimates. The mean values of measured velocity are 27.7, 32.6, and 38.6 mm/s for each different velocity. The corresponding standard deviation of the Doppler frequency estimates were calculated as 0.50, 0.61, and 0.64 Hz, respectively, from the statistical distribution of the estimates for each different velocity measurement with $N = 128$ and $N_{avg} = 200$. The velocity resolution is estimated as 2.7 mm/s by the maximum (worst) standard deviation of 0.64 Hz inferred from (4.22) and substituting it into Eq. 4.8 to convert into velocity.

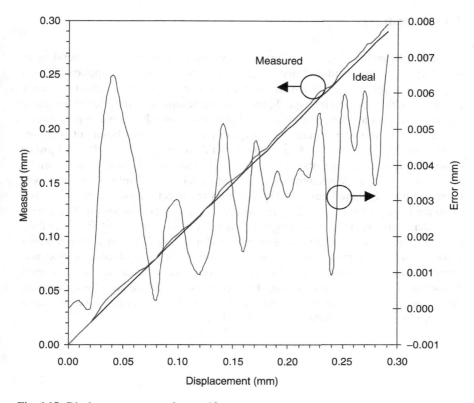

Fig. 4.15 Displacement measured every 10 μm

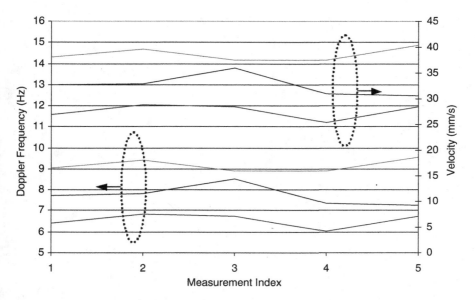

Fig. 4.16 Velocity measurement result for a closing target

4.4 Summary

A double-channel homodyne millimeter-wave interferometric sensor operating at 35.6 GHz was presented and demonstrated for possible sensing applications via displacement sensing and low-velocity measurement. A digital quadrature mixer using a quadrature sampling signal processing technique was introduced for phase detection process to overcome the general problem of non-linear phase response of typically employed conventional analog quadrature mixers. Displacement measurement results indicate that the sensor can detect displacement as small as 10 μm or $\lambda_0/840$, representing a remarkable resolution in terms of wavelength in the millimeter wave range. Measurement error attributed to multiple reflections was corrected by polynomial curve fitting. After error correction, an exceptional maximum measured error of only 27 μm was achieved. Velocity as low as 27.7 mm/s, equivalent to 6.6 Hz in terms of Doppler frequency, was measured at 35.6 GHz for a moving target. The velocity resolution is estimated as 2.7 mm/s. Much lower velocity and better resolution can be measured at the cost of computation and response time. Signal processing for Doppler frequency estimation was developed by means of linear regression on the detected phase combined with the quadrature down-conversion scheme, which provides high resolution and better performance compared to the conventional FFT-based MLE.

Chapter 5
Consideration of Frequency Stability of RF Signal Source for RF Interferometer

Frequency instability of RF signal sources is one of the most important criteria in system design as it affects the accuracy of measurement, particularly the phase measurement due to the highly sensitive nature of the phase. In this chapter, the effects of the frequency instability of RF signal sources on interferometric measurement is analyzed using statistical analysis and FFT spectral estimator. It is shown that the RF source instability has negligible effect on the interferometry-based phase measurement for short time delay.

5.1 Theoretical Analysis of Phase-Noise Effect on Interferometric Measurements

The phase error induced by the instability or phase noise of RF sources affects the accuracy of measurement in general. In this section, the effect of RF source's phase noise on the measurement with RF interferometry is analyzed and investigated using statistical analysis. It is shown that one can predict the stability requirement of the RF source corresponding to the time delay between the transmit and receive signals. It is also demonstrated that the phase noise produces negligible effect on the phase measurement when the time delay is sufficiently small. The result is confirmed with the simulation of a representative millimeter-wave interferometer for displacement measurement.

In an interferometer, the transmit signal from the RF source is mixed with the receive signal (obtained either by reflection from or transmission through the object) in a quadrature mixer to produce a base-band signal. This base-band signal corresponds to the phase difference (due to the time delay) between the transmit and receive signals and can be processed to produce the object information. Interferometry is basically a phase detection process and the measured phase error primarily dictates the accuracy of the target signature. The total induced phase error attributes to the combination of the quadrature mixer imbalance and frequency source

C. Nguyen and S. Kim, *Theory, Analysis and Design of RF Interferometric Sensors*, SpringerBriefs in Physics, DOI 10.1007/978-1-4614-2023-1_5, © Springer Science+Business Media, LLC 2012

instability during the phase detection process. The contribution from the quadrature mixer imbalance can be compensated by correction algorithms in the signal processing as discussed in Chaps. 3 and 4, leaving the instability of the RF source as the principal source of phase error.

The transmit and receive signals in a RF interferometer can be described respectively as

$$v_T(t) = A_T[1 + m(t)] \cos[2\pi f_o t + \phi_n(t) + \phi_i] \tag{5.1}$$

$$v_R(t) = A_R[1 + m(t - t_d)] \cos[2\pi f_o(t - t_d) + \phi_n(t - t_d) + \phi_i] \tag{5.2}$$

where A_T and A_R are the amplitudes of the transmit and receive signals, respectively; $m(t)$ and $\phi_n(t)$ designate the AM (amplitude modulation) and PM (phase modulation) noise effects, respectively; ϕ_i is the initial phase of the frequency source, which can be considered as a constant; and t_d is the time delay between the transmit and receive signals. The effect of the AM noise on the phase detection is generally much smaller than that of the PM noise. Furthermore, in typical balanced quadrature mixers, the AM noise is suppressed by at least 20 dB. Thus, it is reasonable to neglect the AM noise contribution in the estimation of the induced phase error in RF interferometers.

The in-phase and quadrature output signals of the quadrature mixer, after low-pass filtering, are given by

$$v_I(t) = A_I \cos[\Delta\phi_n(t) + \phi_S] + n(t) \tag{5.3}$$

and

$$v_Q(t) = A_Q \cos[\Delta\phi_n(t) + \phi_S] + n(t) \tag{5.4}$$

respectively, where $\Delta\phi_n(t) = \phi_n(t) - \phi_n(t - t_d)$ represents the phase error due to the phase noise of the RF source; $\phi_S = 2\pi f_o t_d$ is the (constant) phase difference between the transmit and receive signals corresponding to the time delay t_d; and $n(t)$ stands for additional noise from the receiver, which can be reduced significantly by averaging the digitized data by data acquisition hardware in the signal processing, making its impact on the system performance negligibly small. The most important contribution to the phase error in the phase detection process is $\Delta\phi_n(t)$. The accuracy of the phase detection is degraded by this noise term, which has uncertainty due to the random nature of noise. It is thus desirable to reduce or eliminate this phase noise effect in measurement. In the following, we will show that this phase noise effect can be neglected when the time delay between the transmit and receive signals is short.

The phase noise $\phi_n(t)$ in a RF source is generally modeled by the Wiener process [51, 52] as a quantity whose time-derivative is zero-mean white Gaussian frequency noise $f_n(t)$. The variance of phase deviation is derived as [46]

$$\sigma^2[\phi_n(\tau)] = E[\phi_n^2(\tau)] - E^2[\phi_n(\tau)] = (2\pi)^2 N_o \tau \tag{5.5}$$

where E denotes the mathematical expectation and N_o is the two-sided frequency noise power spectral density (PSD).

A millimeter-wave source corrupted by phase noise can be represented by

$$v(t) = A \cdot \cos[2\pi f_o t + \phi_n(t) + \theta] \tag{5.6}$$

where A is the maximum amplitude; f_o is the operating frequency; $\phi_n(t)$ is the phase noise; and θ is a random variable, which is uniform over $[0, 2\pi]$ and independent of $\phi_n(t)$. θ makes $v(t)$ stationary so that we can define the autocorrelation of $v(t)$ as

$$R_v(\tau) = E\{v(t)v^*(t+\tau)\} = \frac{A^2}{2}\mathrm{Re}\left\{e^{j2\pi f_o \tau}e^{-\frac{(2\pi)^2}{2}N_o \tau}\right\} \tag{5.7}$$

The power spectral density is obtained as Fourier transform of the autocorrelation, according to Wiener–Khinchin theorem [46], as

$$S_v(\omega) = \int_{-\infty}^{\infty} R_v(\tau)e^{-j\omega\tau}d\tau = \frac{A^2}{2\pi^2 N_o}\left\{1 + \left(\frac{\omega - \omega_o}{2\pi^2 N_o}\right)^2\right\}^{-1} \tag{5.8}$$

where $\omega_o = 2\pi f_o$ is the radian frequency of the source. N_o can be derived from (5.8) as

$$N_o = \frac{\Delta f}{2\pi} \tag{5.9}$$

where Δf is the 3-dB bandwidth of the power spectrum. Substituting (5.9) into (5.5) yields

$$\sigma^2[\phi_n(\tau)] = 2\pi\Delta f \tau \tag{5.10}$$

which implies that the mean square phase deviation is linearly proportional to the time difference τ, as expected for the Wiener process.

As stated earlier, the time derivative of $\phi_n(t)$ is zero mean, white Gaussian frequency noise, which is a wide-sense stationary (WSS) process. Then $\Delta\phi_n(t)$ is also WSS. Thus we can define the autocorrelation of $\Delta\phi_n(t)$ as

$$R_{\Delta\phi_n}(\tau) = E\{\Delta\phi_n(t)\Delta\phi_n(t+\tau)\} = 2R_{\phi_n}(\tau) - R_{\phi_n}(\tau + t_d) - R_{\phi_n}(\tau - t_d) \tag{5.11}$$

The power spectral density corresponding to $R_{\Delta\phi_n}(\tau)$ can be derived using (5.8) as

$$S_{\Delta\phi_n}(\omega) = 2S_{\phi_n}(\omega)(1 - \cos\omega t_d) = t_d^2\omega^2 S_{\phi_n}(\omega)\left(\frac{\sin\omega t_d/2}{\omega t_d/2}\right)^2 \tag{5.12}$$

where $S_{\phi_n}(\omega)$ is the power spectral density for $\phi_n(t)$. The power spectral density corresponding to the derivative of $\phi_n(t)$ is given as [53]

$$S_{\dot{\phi}_n}(\omega) = \omega^2 S_{\phi_n}(\omega) \tag{5.13}$$

One can rewrite (5.12), upon using (5.13), as

$$S_{\Delta\phi_n}(\omega) = t_d^2 S_{\dot{\phi}_n}(\omega)\left(\frac{\sin \omega t_d/2}{\omega t_d/2}\right)^2 \tag{5.14}$$

$\Delta\phi_n(t)$ is a zero mean stationary process, and its variance is computed by

$$\sigma^2[\Delta\phi_n(t)] = R_{\Delta\phi_n}(0) = \frac{t_d^2}{\pi}\int_0^\infty S_{\dot{\phi}_n}(\omega)\left(\frac{\sin \omega t_d/2}{\omega t_d/2}\right)^2 d\omega \tag{5.15}$$

As can be seen from (5.15), the mean square deviation of the phase noise, $\sigma^2[\Delta\phi_n(t)]$, relates to the frequency noise PSD of the frequency source, $S_{\dot{\phi}_n}(\omega)$, which can be measured using a spectrum analyzer. The standard deviation or root mean square (rms) value of $\Delta\phi_n(t)$ is considered as the phase error; i.e. the phase deviation for the delay time t_d due to the phase noise of the frequency source. For the case when $S_{\dot{\phi}_n}(\omega)$ is strictly white noise, (5.15) becomes

$$\sigma^2[\Delta\phi_n(t)] = 2\pi\Delta f\tau \tag{5.16}$$

which is the same as (5.10), after changing t_d to τ without loss of generality. Equation 5.16 allows one to determine the frequency source's stability requirement for a specific time delay to produce negligible phase error. For interferometric measurement with short time delay (e.g., $\tau < 10$ ns), the frequency source instability causes negligible phase error. Note that this conclusion is based on the assumption that the frequency noise PSD is white. Typical frequency noise PSD of a frequency source is not strictly white over the entire frequency range due to the random walk and flicker noise located close to the operating frequency. The assumption of white frequency source PSD, however, is justified when the time delay is short. To illustrate this, we consider a system operating at 36 GHz shown in Fig. 5.1, which represents the operation of a typical homodyne interferometer.

The output voltage of the oscillator representing the millimeter-wave source can be modeled as

$$v_s(t) = V_p \cos\left\{2\pi f_o t + \left[\sum_{m=1}^N \frac{\Delta f_p}{f_m} \sin(2\pi f_m t + \theta)\right]\right\} \tag{5.17}$$

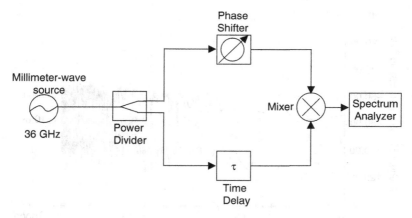

Fig. 5.1 Schematic of a 36-GHz system representing the operation of a typical homodyne interferometer

where V_p is the maximum voltage amplitude; f_m is the offset frequency from the carrier; N represents the number of offset frequency terms used in the phase noise data; Δf_p is the peak frequency deviation resulting from the frequency noise; and θ is uniformly distributed over $[0, 2\pi]$. When the two input signals to the mixer are quadrature in phase, the output signal of the mixer at an offset frequency can be derived from Fig. 5.1 as

$$v_o(t) = K \sin\left(\frac{2\Delta f_p}{f_m}\right) \sin(\pi f_m \tau) \cos\left\{2\pi f_m\left[t - \left(\frac{\tau}{2}\right)\right]\right\} \qquad (5.18)$$

where K measures the change of the mixer's output voltage due to change in the frequency noise. For the typical case of $\Delta f_p < < f_m$ and when $f_m < < \frac{1}{\tau}$ (for short time delay), (5.18) reduces to

$$v_o(t) = 2\pi K \Delta f_p \tau \cos\left\{2\pi f_m\left[t - \left(\frac{\tau}{2}\right)\right]\right\} \qquad (5.19)$$

which implies that the output voltage from the mixer at an offset frequency is linearly proportional to the frequency noise. Also, (5.19) shows that the frequency noise may produce negligible voltage from the mixer when the time delay is sufficiently short. To confirm this phenomenon, we show in Fig. 5.2 the frequency noise PSD at the mixer's output as a function of offset frequency for different delay times calculated by the Agilent ADS simulator [54]. Typical phase noise of a Ka-band oscillator was used for the 36-GHz millimeter-wave source. The power divider, mixer, phase shifter and delay line were ideal components taken from the Agilent ADS library. As can be seen, the shorter the time delay, the greater suppression of the frequency noise components at the low frequency offset in base-band. It is therefore reasonable to approximate that $S_{\dot\phi_n}(\omega)$ is white

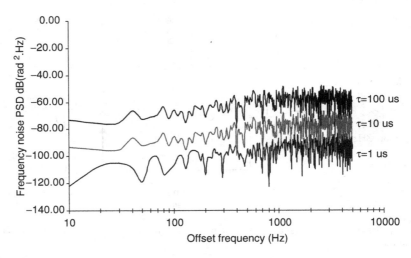

Fig. 5.2   Calculated time-delay effect on the frequency noise PSD

for short time delay. As an example, we consider a millimeter-wave interferometric sensor shown in Fig. 3.1 for displacement measurement. The range from the antenna to the target for normal incidence of wave relates to the detected phase difference $\phi(t)$ as

$$r(t) = \frac{\phi(t)}{4\pi} \lambda_0 \tag{5.20}$$

where λ_0 is the operating wavelength in air. The range variation is produced by changes in the surface profile and can be represented in the time domain as

$$r(nT) = r(nT) - r[(n-1)T] \qquad n = 1,\ 2,\ 3,\ \cdots \tag{5.21}$$

where T is the sampling interval. The displacement can be described as a summation of consecutive range variations:

$$d(nT) = \sum_{n=1}^{k} r(nT) \qquad n = 1,\ 2,\ 3,\ \cdots,\ k \tag{5.22}$$

Error in phase measurement thus produces displacement error through (5.20). Figure 5.3 shows the rms value of the displacement error at 36 GHz as a function of time delay for various 3-dB bandwidths of the frequency source. From Fig. 5.3, we can deduce that the phase noise of the frequency source has little impact on the coherent phase detection using the interferometer for short time delay.

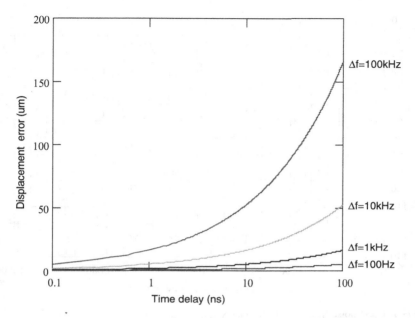

Fig. 5.3 Rms displacement error of a millimeter-wave interferometer operating at 36 GHz

5.2   Phase Noise Estimation

Frequency instability of a RF signal source contributes to the error that affects measurement accuracy. It is typically characterized by a phase noise spectrum in the frequency domain or Allan variance in the time domain [55]. In the presented double-channel homodyne millimeter-wave interferometric sensor, the phase noise of the millimeter-wave signal is down-converted and appears in the second IF signal, whose frequency is low enough to be manipulated by digital signal processing. We estimated the phase noise of the signal source through a phase noise spectrum obtained by the FFT spectral estimator, as shown in Fig. 5.4, for the actual measurement- and reference-channel signals. In the phase noise spectrum, the sampling frequency was chosen as 200 kHz with $N = 4,096$, and a Hanning window was used. Each spectrum was measured 50 times and then averaged.

A signal containing amplitude and phase noise can be represented, neglecting the inter-modulation of the amplitude and phase noise due to nonlinearity of the signal source or the components following it, as

$$v(t) = \left[A_p + a(t)\right]\left[\cos 2\pi f_c t + \phi_n(t)\right] \tag{5.23}$$

where $a(t)$ and $\phi_n(t)$ denote the respective amplitude and phase noise, and A_p and f_c are the amplitude and frequency of the signal, respectively. The double side-band power spectrum of (5.23) is a superposition of the carrier power and power spectrum of the amplitude and phase noise. That is,

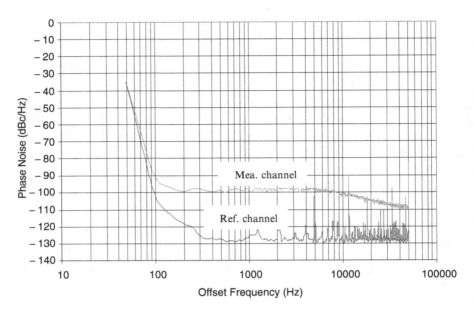

Fig. 5.4 Phase-noise spectrum of each channel

$$S(f) = \frac{A_p^2}{2}\delta(f - f_c) + S_a(f - f_c) + S_\phi(f - f_c) \tag{5.24}$$

where δ is the Dirac delta function, and the second and third term represent the amplitude and phase noise power spectra, respectively. For a signal where external additive noise is predominant, as assumed for the reference-channel signal in our system, the power spectrum has equal contribution from the amplitude and phase noise [56, 57]. The phase noise spectrum is thus 3 dB below the power spectrum normalized to the carrier power in a 1-Hz bandwidth. For the measurement-channel signal, the noise components of the signal are down-converted from the RF signal source and governed by it. The spectrum of a RF signal is typically dominated by the phase (or frequency) noise for the frequencies close to the carrier, and by amplitude noise for the frequencies more than a few tens of kilohertz from the carrier [58]. From this rationale, we can then neglect the contribution of the amplitude noise for the measurement-channel signal because we are interested only in the frequencies close to the carrier for phase noise estimation. If we model the sinusoidal signal corrupted by phase noise as a linear frequency modulation process, then the signal is described as

$$v(t) = A_P \cos[2\pi f_c t + \eta(f_i)\sin(2\pi f_i t)] \tag{5.25}$$

where f_i is the modulating frequency, and $\eta(f_i)$ is the modulation index defined by

$$\eta(f_i) = \frac{\delta f}{f_i} \tag{5.26}$$

with δf being the peak frequency deviation at the modulating frequency f_i due to the frequency instability of the signal source. The phase noise spectrum of the signal can be approximated as power spectral density normalized to carrier power when the amplitude noise contribution is negligible. The single sideband (SSB) phase noise power $P_{ssb}(f_i)$ at an offset (or modulating) frequency f_i is related to the signal power, P_c, and the root mean square of the modulation index, η_{rms}, as

$$\frac{P_{ssb}(f_i)}{P_c} = \frac{\eta_{rms}^2(f_i)}{2} \qquad (5.27)$$

The SSB phase noise, in dBc/Hz, can be obtained by

$$L(f_i) = P_{ssb}(f_i) - RBW - P_C \qquad (5.28)$$

where RBW represents the resolution bandwidth in dB, and the power is measured in dBm. Then the integrated phase noise variance is expressed as

$$\overline{\sigma_\phi^2} = 2 \int_{f_L}^{f_H} L(f_i)df_i rad^2 (dB) \qquad (5.29)$$

where the bar denotes statistical average, f_H is defined as the upper-band limit of the band-limited differential amplifier of the sensor, and f_L is determined from the total observation time, NT, as

$$f_L = \frac{1}{2NT} \qquad (5.30)$$

with N being the number of data points and T being the sampling time.

On the basis of the fact that frequency is time derivative of phase given as

$$f(t) = \frac{1}{2\pi} \frac{d\phi(t)}{dt} \qquad (5.31)$$

the power spectrum pair of the frequency and phase functions has the following relationship:

$$S_f(f) = (2\pi f)^2 S_\phi(f) \qquad (5.32)$$

The variance of frequency noise over the same bandwidth can be obtained, making use of (5.29) and (5.32), as

$$\overline{\sigma_f^2} = 2 \int_{f_L}^{f_H} L(f_i)f_i^2 df_i \qquad Hz^2(dB) \qquad (5.33)$$

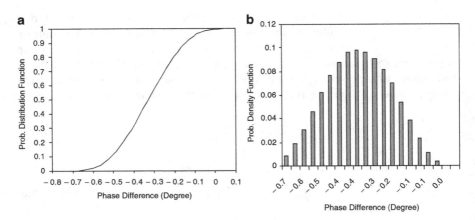

Fig. 5.5 Probability distribution function (**a**) and probability density function (**b**) for the phase difference

If the phase noise is white over the bandwidth, with the frequency noise spectrum having a rising slope of 20 dB/decade with increasing frequency, then (5.29) can be simplified to

$$\overline{\sigma_\phi^2} = 2[L(f_i)(f_H - f_L)] \qquad rad^2(dB) \tag{5.34}$$

and (5.33) reduces to

$$\overline{\sigma_f^2} = 2\left[L(f_i)\left(\frac{f_H^3 - f_L^3}{3}\right)\right] \qquad Hz^2(dB) \tag{5.35}$$

The probability distribution and density functions of the phase difference defined by (4.13) are illustrated in Fig. 5.5, which shows the characteristics of Gaussian distribution. These results demonstrate that the phase noise process can be approximated as white Gaussian noise process over the bandwidth of (5.29). They were obtained by sampling 2,000 data points and averaging 50 times.

Therefore, it is possible to determine the root mean square (rms) phase and frequency error originated from the instability of the frequency source by (5.34) and (5.35), respectively. The rms phase error was estimated as 0.13° for the measurement-channel signal, by substituting $f_H = 9.81$ kHz and $f_L = 100$ Hz into (5.34). For the reference channel, the error was obtained as 0.004°. The contribution of the frequency-source instability coming from DDS to phase error is therefore negligible. The rms frequency error estimated by (5.35) was 12.6 Hz for the same f_H with $f_L = 781$ Hz.

Chapter 6
Summary, Conclusion and Applications

In the previous chapters, the theory, analysis and design of RF interferometric sensors have been presented for sensing applications. Two types of system architecture were used to implement the millimeter-wave interferometric sensors: homodyne and double-channel homodyne. These sensors were realized on planar structures using MMIC and MIC technology for light, compact, and low-cost design in addition to some external components (phase-locked oscillator, directional coupler, and horn antenna.)

The homodyne configuration can be generally accepted for most sensing applications. Results of the displacement measurement using the developed homodyne millimeter-wave interferometric sensor show that sub-millimeter resolution in the order of 0.05 mm is feasible without correcting the non-linear phase response of the quadrature mixer used in the sensor.

The double-channel homodyne configuration makes it possible to exclude the non-linearity of the quadrature mixer and to obtain better resolution than its homodyne counterpart with only a slight increase in circuit complexity. The designed double-channel homodyne millimeter-wave interferometric sensor provides a remarkable resolution of only 0.01 mm or 1/840th of the operating wavelength. In this sensor, a digital quadrature mixer, constituted by a quadrature-sampling signal processing technique, was implemented to avoid the nonlinear phase response of conventional quadrature mixers typically employed. The sensor demonstrates that displacement sensing with micron resolution and accuracy and high-resolution low-velocity measurement are feasible using millimeter-wave RF interferometry, which is attractive not only for displacement and velocity measurement, but also for other sensing applications requiring very fine resolution and accuracy.

RF interferometric sensors, particularly those in the millimeter-wave frequency region of RF, have a strong potential for sensing applications requiring fine resolution approaching sub-millimeter and fast response. They are capable to replace laser interferometers used in harsh environments with dust, smoke, fog, and even danger of explosion. Although only single-frequency RF interferometry based on the homodyne architecture is considered and only displacement, liquid-level, and velocity sensing are demonstrated in this book, the system architectures, numbers

C. Nguyen and S. Kim, *Theory, Analysis and Design of RF Interferometric Sensors*,   65
SpringerBriefs in Physics, DOI 10.1007/978-1-4614-2023-1_6,
© Springer Science+Business Media, LLC 2012

of frequency involved, and sensing applications are not limited with these. With the advances of RF technology, from low to extremely high frequencies, in solid-state devices, circuits and systems, coupled with many existing sensing possibilities and demands for new sensing applications, it is envisioned that novel RF interferometric sensors is continued to be developed and implemented for vast sensing needs. One possibility is development of extremely small and low-cost RF interferometric sensors integrated completely on single CMOS/BiCMOS chips that can be deployed in large networks for sensing and imaging. It is noted that sensing applications indeed drive the development of sensors in general and RF interferometers in particular, and in turn RF interferometers, once developed, make it possible for new applications to be realized and pushing the technology further.

From the point of view of system architectures, different RF interferometric architectures can be evolved which may be used for new applications or benefit existing applications. Some of these systems may be adapted from the optical interferometry from which the RF interferometry was originated. For instance, a RF interferometric sensor can be configured with RF source of two frequencies for absolute distance measurement or ranging, which is an indispensable requirement of industrial sensing, in addition to velocity measurement. Another is possible use of RF interferometry to overcome the drawback of conventional pulse or FMCW radar in short-distance ranging. Typically, it is difficult to use these conventional techniques in short-distance ranging because of relatively poor resolution limited by the signal's bandwidth. These techniques require very short pulses or wide bandwidths, which make it difficult to design system's components, besides the difficulty in resolving very short time delay caused by short distances. Combining RF interferometry with FMCW or pulse technique for sensors promises to produce an attractive solution to compensate for the limitation in short-distance applications, exploiting the high range resolvability or phase resolvability of RF interferometry.

The RF interferometry can be extended to radar sensors for automobile, which has a huge potential market. Millimeter-wave radar has been intensively investigated for intelligent cruise control and safety for vehicles, and most radar architectures have adopted the FMCW technique. The RF interferometry combined with the FMCW technique at millimeter wavelengths promises to increase the resolution of the radar sensors by the phase-sensitive time-domain signal processing in addition to FFT, frequency-domain signal processing; and, above all, to solve the conventional short-distance ranging problem of FMCW and pulse radar.

Near-field imaging is also one possible interesting application of the RF interferometric sensors discussed in this book. Non-destructive material testing and microscopy can be categorized as those of near-field imaging. In near-field imaging using probes, the resolution especially relies on the probe dimension, such as the aperture size of an open-ended waveguide probe or the conductor diameter of an open-ended coaxial cable probe, rather than the operating frequency. Thus, by miniaturizing the probe, which is feasible with the advances of technology, it is possible to achieve high resolution at relatively low operating frequencies.

The presented RF interferometric sensors may be used for medical and biological purposes, such as detecting vital signals from the hearts or lungs of human or animals through the detected phase. Further extension of this application that can be envisioned is remote diagnoses by installing a RF interferometric sensor in mobile phones and connecting it through wireless networks, so that vital signals detected can be transferred to authorized healthcare personnel in remote locations.

Finally, we now wish to restate one implicit principle of sensing: sensing involves detecting changes and any change causes variation of the phase of electrical signals used for sensing, and so any technique that can detect quickly and accurately the phase change would serve as a good candidate for sensing. RF interferometry is such an attractive sensing technique. Recognizing this sensing principle and the strength of the RF interferometric sensing in detecting accurately the phase, we can indeed envision vast applications for this technology. Some of these have been discussed earlier and some other possibilities are: remote measurement and monitoring of the displacement or stability of man-made and natural structures such as buildings, dams, bridges, towers, industrial plants, mountain slopes, ocean waves, etc.; monitoring of volcanoes and seismic activity; detecting seismic waves; locating buried mines and unexploded ordnance (UXO); examination and monitoring of civil infrastructures such as pavements, bridges, power plants, etc.; personnel health care examination and monitoring; etc.

References

1. A. Kolodziejczyk, M. Sypek, A.A. Michelson-life and achievements. Proc SPIE **1121**, 2–14 (1989)
2. J.M. Schmitt, Optical coherence tomography (OCT): a review. IEEE J. Slect. Top. Quant. Electron. **5**(4), 1205–1215 (1999)
3. M.M. Gualini, W.A. Kha, W. Sixt, H. Steinbichler, Recent advancements of optical interferometry applied to medicine, in *Multi Topic Conference, 2001. IEEE INMIC 2001. Technology for the 21st Century. Proceedings. IEEE International*, 28–30 Dec 2001, pp. 205–211
4. W.H. Steel, Another look at the theory of interferometer. Proc SPIE **1121**, 18–23 (1989)
5. W.H. Steel, *Interferometry* (Cambridege University, London, 1983)
6. R.J. King, *Microwave Homodyne Systems* (Peter Peregrinus, London, 1978)
7. R. Zoughi, *Microwave Non-Destructive Testing and Evaluation* (Kluwer Academic, Boston, 2000)
8. M.A. Heald, C.B. Wharton, *Plasma Diagnostics with Microwaves* (John Wiley, New York, 1965)
9. A. Stelzer, C.G. Diskus, K. Lubke, H.W. Thim, Microwave position sensor with submillimeter accuracy. IEEE Trans. Microw Theory Tech. **47**(12), 2621–2624 (1999)
10. A. Benlarbi-Delai, D. Matton, Y. Leroy, Short-range two-dimension positioning by microwave cellular telemetry. IEEE Trans. Microw Theory Tech. **42**(11), 2056–2062 (1994)
11. A. Benlarbi, J.C. Van De Velde, D. Matton, Y. Leroy, Position, velocity profile measurement of a moving body by microwave interferometry. IEEE Trans. Instrum. Meas **39**(4), 632–636 (1990)
12. P.E. Engler, S.S. Reisman, C.Y. Ho, A microwave interferometer as a non contacting cardiopulmonary monitor, in *1988 Bioengineering Conference, Proceedings of the 1988 Fourteenth Annual Northeast*, Durham, 10–11 Mar 1988, pp. 62–65
13. A.R. Thompson, J.M. Moran, G.W. Swenson Jr., *Interferometry and Synthesis in Radio Astronomy* (John Wiley, New York, 1986)
14. W. B. Doriese, A 145-GHz Interferometer for Measuring the Anisotropy of the Cosmic Microwave Background, Ph.D. Dissertation, Physics Department. Princeton University, Princeton, 2002
15. G. Franceschetti, R. Lanari, *Synthetic Aperture Radar Processing* (CRC press, New York, 1999). Ch. 4
16. W.F. Feltz, H.B. Howell, R.O. Knuteson, H.M. Woolf, H.E. Revercomb, Near continuous profiling of temperature, moisture, and atmospheric stability using the atmospheric emitted radiance interferometer (AERI). J. Appl. Meteor. **42**, 584–597 (2003)
17. E.N. Ivanov, M.E. Tobar, R.A. Woode, Microwave interferometry: application to precision measurements and noise reduction techniques. IEEE Trans. Ultrason., Ferroelect., Freq. Contr **45**(6), 1526–1536 (1998)

C. Nguyen and S. Kim, *Theory, Analysis and Design of RF Interferometric Sensors*,
SpringerBriefs in Physics, DOI 10.1007/978-1-4614-2023-1,
© Springer Science+Business Media, LLC 2012

18. S. Kim, C. Nguyen, A displacement measurement technique using millimeter-wave interferometry. IEEE Trans. Microw Theory Tech. **51**(6), 1724–1728 (2003)

19. S.T. Kim, C. Nguyen, On the development of a multifunction millimeter-wave sensor for displacement sensing and low-velocity measurement, *IEEE Trans. on microwave theory and Techniques*, Vol. MTT-52, No. 11, Nov. 2004, pp. 2503–2512

20. J. Musil, F. Zacek, *Microwave Measurements of Complex Permittivity by Free Space Methods and Their Application* (Elsevier, Amsterdam, 1986). Ch. 4

21. C.W. Domier, W.A. Peebles, N.C. Luhmann, Millimeter-wave interferometer for measuring plasma electron density. Rev. Sci. Instrum. **59**(8), 1588–1590 (1988)

22. X. Liu, W. Clegg, D.F.L. Jenkins, B. Liu, Polarization interferometer for measuring small displacement. IEEE Trans. Instrum. Meas **50**(4), 868–871 (2001)

23. M. Norgia, S. Donati, D. D'Alessandro, Interferometric measurements of displacement on a diffusing target by a speckle tracking technique. IEEE J. Quant Electron. **QE-37**(6), 800–806 (2001)

24. B.G. Zagar, A laser-interferometer measuring displacement with nanometer resolution. IEEE Trans. Instrum. Meas **43**(2), 332–336 (1994)

25. K. Itoh, Analysis of the phase unwrapping problem. Appl. Opt. **21**(14), 2470 (1982)

26. D.C. Ghiglia, M.D. Pritt, *Two-Dimensional Phase Unwrapping Theory, Algorithms, and Software* (John Wiley, New York, 1998). Ch. 1

27. A.V. Oppenheim, R.W. Schafer, *Digital Signal Processing* (Prentice-Hall, Englewood Cliffs, NJ, 1975). Ch.10.6

28. *IE3D*, Zeland Software, Inc., Fremont, California

29. S.J. Goldman, *Phase Noise Analysis in Radar Systems Using Personal Computers* (John Wiley, New York, 1989). Ch. 10

30. F.E. Churchill, G.W. Ogar, B.J. Thompson, The correction of I and Q errors in a coherent processors. IEEE Trans. Aerosp. Electron. Syst **AES-17**(1), 131–137 (1981)

31. D.E. Noon, Wide band quadrature error correction (using SVD) for stepped-frequency radar receivers. IEEE Trans. Aerosp. Electron. Syst **AES-35**(6), 1444–1449 (1999)

32. R.A. Monzingo, S.P. Au, Evaluation of image response signal power resulting from I-Q channel imbalance. IEEE Trans. Aerosp. Electron. Syst **AES-23**(2), 285–287 (1987)

33. *33-43GHz GaAs MMIC Image Rejection Balanced Mixer*, AM038R1-00, Data Sheet, Alpha Industries, Inc., Woburn, MA

34. J. Otto, Radar applications in level measurement, distance measurement and nondestructive material testing, in *Microwave Conference and Exhibition, 27th European*, vol. 2, Sep 1997, pp. 1113–1121

35. H.H. Meinel, Commercial applications of millimeter waves history, present status, and future trends. IEEE Trans. Microw Theory Tech. **43**(7), 1639–1653 (1995)

36. M. Wollitzer, J. Buechler, J.F. Luy, U. Siart, E. Schmidhammer, J. Detlefsen, M. Esslinger, Multifunctional radar sensor for automotive application. IEEE Trans. Microw Theory Tech. **46**(5), 701–708 (1998)

37. R.H. Rasshofer, E.M. Biebl, Advanced millimeterwave speed sensing system based on low-cost active integrated antennas. IEEE MTT-S Int. Microw Symp. Dig. **1**, 285–288 (1999)

38. I. Gresham, N. Jain, T. Budka, A. Alexanian, N. Kinayman, B. Ziegner, S. Brown, P. Staecker, A compact manufacturable 76-77-GHz radar module for commercial ACC applications. IEEE Trans. Microw Theory Tech. **49**(1), 44–58 (2001)

39. N. Weber, S. Moedl, M. Hackner, A novel signal processing approach for microwave Doppler speed sensing. IEEE MTT-S Int. Microw Symp. Dig. **3**, 2233–2235 (2002)

40. F. Xiao, F.M. Ghannouchi, T. Yakabe, Application of a six-port wave-correlator for a very low velocity measurement using the Doppler effect. IEEE Trans. Instrum. Meas **52**(2), 297–301 (2003)

41. W.M. Waters, B.R. Jarret, Bandpass signal sampling and coherent detection. IEEE Trans. Aerosp. Electron. Syst **AES-18**, 731–736 (1982)

42. D.W. Rice, K.H. Wu, Quadrature sampling with high dynamic range. IEEE Trans. Aerosp. Electron. Syst **AES-18**, 736–739 (1982)
43. V. Considine, Digital complex sampling. Electron Lett **19**(16), 608–609 (1983)
44. C.R. Rader, A simple method for sampling in-phase and quadrature composition. IEEE Trans, Aerosp. Electron. Syst. **20**(6), 821–824 (1984)
45. H. Liu, A. Ghafoor, P.H. Stockmann, A new quadruture sampling and processing approach. IEEE Trans. Aerosp. Electron. Syst **AES-25**(5), 733–748 (1989)
46. A. Papoulis, *Probability, Random Variables, and Stochastic Processes* (McGraw-Hill, New York, 1984)
47. G.A.F. Seber, A.J. Lee, *Linear Regression Analysis* (John Wiley, New York, 2003)
48. S.A. Tretter, Estimating the frequency of a noisy sinusoid by linear regression. IEEE Trans. Inform. Theory **IT-31**, 832–835 (1985)
49. C.D. Cain, A. Yardim, E.T. Katsaros, Performance of an FIR filter-based spectral centroid tracker for Doppler determination. IEEE Int Symp Circ Syst **5**, 2455–2458 (1991)
50. G.H. Golub, C.F. Van Loan, *Matrix Computation* (The John Hopkins University, Baltimore, MD, 1989). Ch. 8
51. J. Salz, Coherent lightwave communications. AT & T Tech J. **64**(10), 2153–2209 (1985)
52. J.R. Barry, E.A. Lee, Performance of coherent optical receivers. Proc. IEEE **78**(8), 1369–1394 (1990)
53. H. Taub, D.L. Schilling, *Principles of Communication Systems* (McGraw-Hill, New York, 1971)
54. *Advanced Design System (ADS)*, Agilent Technologies Inc., Santa Clara, CA
55. J.A. Barnes, A.R. Chi, L.S. Cutler, D.J. Healey, D.B. Leeson, T.E. Mcgunigal, J.A. Mullen Jr., W.L. Smith, R.L. Sydnor, R.F.C. Vessot, G.M.R. Winkler, Characterization of frequency stability. IEEE Trans. Instrum. Meas. **20**(2), 105–120 (1971)
56. W.P. Robins, *Phase Noise in Signal Sources* (Peter Peregrinus, London, 1998). Ch. 3
57. J. Rutman, Characterization of phase and frequency instability in precision frequency sources: fifteen years of progress. Proc. IEEE **66**(9), 1048–1075 (1978)
58. J.R. Ashley, T.A. Barley, G.J. Rast, The measurement of noise in microwave transmitters. IEEE Trans. Microw Theory Tech. **25**(4), 294–318 (1977)

Index

A

Amplitude and phase imbalance, 26, 27, 29, 31
Amplitude imbalance, 26–28, 31, 32
Analog quadrature mixers, 54

B

Boundary conditions, 8

C

Center of gravity, 46
Conductivity, 9
Conventional short-distance ranging
    problem, 66

D

DFT. *See* Discrete fourier transform
Dielectric constant, 10
    of free space, 9
Dielectric loss tangent, 10
Digital converter (DC) offset, 13, 17, 23, 26,
    27, 29, 31, 41
Digital quadrature mixer (DQM), 14, 36, 38,
    41, 42, 44, 54, 65
Dirac delta function, 62
Discrete fourier transform (DFT), 29
Displacements, 15–17, 20, 23–25, 33, 35, 36,
    38–40, 42, 48–51, 54, 55, 60, 61, 65
    measurement, 15
Doppler
    frequency, 35–36, 38–40, 42–46, 52, 54
    radar, 35
    velocimeter, 36
    velocimetry, 38
    velocity, 35, 36

Double-channel heterodyne, 44, 46, 48
    interferometer, 41
Double-channel homodyne, 4, 5, 35, 36, 54,
    61, 65
Double-channel system, 3
DQM. *See* Digital quadrature mixer

E

Electric fields, 8, 9
Electromagnetic wave, 8

F

Fast fourier transform (FFT), 35, 36, 44–46, 54
    spectral estimator, 39, 61–62
FFT. *See* Fast fourier transform
FMCW. *See* Frequency modulated
    continuous wave
$1/f$ noise, 4, 13
Fourier transform, 27
Frequency modulated continuous wave
    (FMCW), 5, 66

G

Gram-Schmidt orthogonalization, 29

H

Hanning window, 62
Hermitian image, 28
Heterodyne
    interferometer, 4
Homodyne, 4, 5, 13, 15, 20, 22, 25, 27,
    31, 33, 35, 65
    interferometer, 58

I

Image-to-signal ratio (ISR), 27, 28, 31, 32
In-phase, 16, 25, 31, 41
    and quadrature, 13
Instability, 31, 55, 58, 61, 64
Interferometer, 1, 3, 4, 15, 35, 36, 55, 61
Interferometric sensors, 4, 15, 20, 22, 25, 33,
        35, 44, 46, 48, 54, 60, 61, 65–67
Interferometry, 1, 8, 17, 55
Intrinsic impedances, 8, 11, 12
I/Q error, 4, 15, 25, 26, 28–29, 31, 33
ISR. *See* Image-to-signal ratio

L

Linear regression, 45, 46, 54
Liquid-level gauging, 24, 33, 35

M

Mach-Zehnder interferometer, 2
Maximum likelihood estimation (MLE),
        45, 46, 54
Mean square error (MSE), 50
Michelson, A.A, 1
Michelson interferometer, 1
Microwave and millimeter-wave integrated
        circuits, 15
Microwave integrated circuits (MICs), 5, 15,
        20, 33, 36, 46, 65
Microwave monolithic integrated circuits
        (MMICs), 5, 15, 20, 21, 33, 36, 46, 65
MICs. *See* Microwave integrated circuits
Millimeter-wave frequencies, 4
Millimeter-wave interferometer, 15
MLE. *See* Maximum likelihood estimation
MMICs. *See* Microwave monolithic
        integrated circuits
Monostatic system, 3
MSE. *See* Mean square error

N

Non-destructive characterization, 15

O

Optical interferometers, 15
Optical interferometry, 1, 20

P

Phase and amplitude imbalances, 41
Phase imbalance, 26, 28, 31, 32
Phase noise, 13, 35, 38, 39, 43, 55, 56, 58, 61–64

Phase shifter, 13
Phase unwrapping, 17–20, 23–24, 42, 43
Plasma diagnostics, 15
Propagation constant, 8
Pulse, 66

Q

Quadrature, 16, 25, 31, 41
    mixers, 3, 4, 13–18, 23, 25–28, 31–33, 35,
        36, 38, 41, 55, 56, 65
Quadrature sampling, 38, 41
Quadrature sampling digital signal processing,
        14, 35
Quadrature sampling signal processing, 38, 41,
        54, 65
Quadrature upconverter, 3

R

Radar velocimetry, 39
Radio frequency (RF)
    interferometric sensors, 5
    range, 1
    sensors, 4
Radio frequency (RF) interferometers, 2–5, 8,
        9, 12–18, 25, 28, 35, 55, 56, 65–67
Reflection, 8
    coefficient, 11, 12
    and transmission coefficient, 12
    and transmission method, 12
Reflection/transmission coefficient, 13
Reflectometer, 3
Relative dielectric constant, 9–12
Resolution, 15, 25, 28, 33, 35, 36, 43–45, 51,
        54, 65, 66
RF. *See* Radio frequency

S

Sampling interval, 17, 61
Sampling time, 14, 29, 39
Signal processing, 66
Singular value decomposition (SVD), 50–51
Standard deviation, 52
SVD. *See* Singular value decomposition

T

Transmission coefficients, 8, 12

V

Velocity, 8, 35, 36, 38, 40, 43, 51, 52, 54, 65, 66
    resolution, 54